# NORTHROP YF-17 COBRA

## A Pictorial History

### Don Logan

Schiffer Military/Aviation History
Atglen, PA

## ACKNOWLEDGMENTS

I would like to thank the following individuals who have helped me in this project: Mr. C. John Amrhein of Northrop/Grumman Corporation, D.J. Fisher, Michael A. France, Marty Isham, Dennis Jenkins, Roger Johansen, Craig Kaston, Robert L. Lawson, Roy Lock, Brian C. Rogers, and Mick Roth.

## THE AUTHOR

After graduating from California State University-Northridge with a BA degree in History, Don Logan joined the USAF in August of 1969. He flew as an F-4E Weapon Systems Officer (WSO), stationed at Korat RTAFB in Thailand, flying 133 combat missions over North Vietnam, South Vietnam, and Laos before being shot down over North Vietnam on July 5, 1972. He spent nine months as a POW in Hanoi North Vietnam. As a result of missions flown in Southeast Asia, he received the Distinguished Flying Cross, the Air Medal with twelve oak leaf clusters, and the Purple Heart. After his return to the U.S., he was assigned to Nellis AFB where he flew as a rightseater in the F-111A. He left the Air Force at the end of February 1977.

In March of 1977 Don went to work for North American Aircraft Division of Rockwell International, in Los Angeles, as a Flight Manual writer on the B-1A program. He was later made Editor of the Flight Manuals for B-1A #3 and B-1A #4. Following the cancellation of the B-1A production, he went to work for Northrop Aircraft as a fire control and ECM systems maintenance manual writer on the F-5 program.

In October of 1978 he started his employment at Boeing in Wichita, Kansas as a Flight Manual/Weapon Delivery manual writer on the B-52 OAS/CMI (Offensive Avionics System/Cruise Missile Integration) program. He is presently the editor for Boeing's B-52 Flight and Weapon Delivery manuals, B-1 OSO/DSO Flight Manuals and Weapon Delivery Manuals, and T-43A flight Manuals.

Don Logan is also the author of *Rockwell B-1B: SAC's Last Bomber*, *The 388th Tactical Fighter Wing: At Korat Royal Thai Air Force Base 1972*, and *Northrop's T-38 Talon: A Pictorial History* (all three titles are available from Schiffer Publishing Ltd.).

Book Design by Robert Biondi.

Copyright © 1996 by Don Logan.
Library of Congress Catalog Number: 95-71769

All rights reserved. No part of this work may be reproduced or used in any forms or by any means – graphic, electronic or mechanical, including photocopying or information storage and retrieval systems – without written permission from the copyright holder.

Printed in China.
ISBN: 0-88740-910-5

We are interested in hearing from authors with book ideas on related topics.

---

Published by Schiffer Publishing Ltd.
77 Lower Valley Road
Atglen, PA 19310
Please write for a free catalog.
This book may be purchased from the publisher.
Please include $2.95 postage.
Try your bookstore first.

# INTRODUCTION

The Northrop YF-17 holds a special place in aircraft history. The YF-17 was one of the two prototypes tested in the U.S. Air Force Air Combat Fighter competition, a program which attempted to reverse the trend of increasing cost and complexity of new fighter aircraft, and which resulted in the selection and manufacture of the F-16 as the next generation free world fighter. Even though the YF-17 lost the USAF competition, it was the prototype for the U.S. Navy's F/A-18 aircraft.

The YF-17 was the culmination of a line of Northrop light weight fighter designs starting with the N-102 FANG in 1956 continuing through the N-156 Freedom Fighter series (later designated F-5 by the U.S. government), ending with the P530 and P600 Cobra series (later designated the YF-17).

The F/A-18, designed and produced by Northrop and McDonnell Douglas was the first modern aircraft resulting from a design/production team comprised of U.S. aerospace firms. This trend has become the industry standard and continued with the ATF (Advanced Tactical Fighter) program which was made up of two design teams, Boeing, General Dynamics, and Lockheed for the YF-22, and Northrop and McDonnell Douglas for the YF-23.

The two prototype YF-17 aircraft, (72-1569 and 72-1570), started a line of F-18s in service in the U.S. Navy and Marine Corps, and flown by other countries including Australia, Canada, Finland, Kuwait, South Korea, Spain, and Switzerland.

The two Northrop YF-17s sit on the Edwards AFB ramp at sunrise ready for a day of testing. (Northrop)

**The N-102 FANG was classified as a light weight day fighter and was designed to be easily converted into a fighter-bomber. The highly swept delta wing and tail surfaces of the mockup suggest the Mach 2 capabilities of the design. (Northrop)**

**The FANG was a single engine design with a high wing. The aircraft was designed for easy maintenance and weapons loading. Of interest is the underslung engine air intake, similar to the F-16. (Northrop)**

# LIGHT WEIGHT FIGHTER PROGRAM HISTORY

The light weight fighter program was developed in response to a need for a new fighter for NATO and other U.S. allies and, as a result of the war in Southeast Asia. The newest fighter in the USAF inventory was the Republic F-105 Thunderchief fighter-bomber. There were new tactical fighters available for production to fill this requirement. The General Dynamics F-111 (TFX) was entering production, but the F-111, like the F-105, was a fighter-bomber with very limited air-to-air capability. As a result, the aircraft currently in production which could be adapted for use by the Air Force as tactical fighters were two U.S. Navy designs, the McDonnell F-4 Phantom II, and the Vought (Ling Temco Vought) A-7 Corsair II. Two other U.S. designed tactical fighter aircraft, the Lockheed F-104 Starfighter and the Northrop F-5 Freedom Fighter, were also in production during the mid-1960s. These two types saw limited use by the USAF but were quite successful in foreign military service. Only about 300 F-104s had been procured by the U.S. Air Force in the late 1950s, and were used as point defense interceptor. By the early 1960s, these aircraft were already being phased out of the active Air Force inventory.

Lockheed continued development of the F-104 resulting in a heavier, multi-role version designated the F-104G. The G model was sold through a co-production agreement to four European nations: Germany, Italy, Belgium, and The Netherlands. Co-production of the fighter also occurred in both Canada and Japan. Approximately 2,500 F-104 series aircraft were produced and 15 countries operated F-104s. As a possible replacement for the F-104 in the USAF inventory, Northrop was awarded a contract to produce the F-5 Freedom Fighter based on Northrop's N-156 design. This airplane was a light weight (less than 13,000 pounds), twin-engine tactical fighter developed from the same Northrop design series as the T-38 Talon supersonic jet trainer. The T-38 had entered USAF service as their advanced pilot training aircraft in 1961. The highly successful T-38 became the first trainer aircraft to set the jet aircraft world absolute time-to-climb speed records.

The F-5 aircraft series proved to be a relatively inexpensive, simple, capable fighter which could be supplied to countries which did not have the expertise or budget to operate more complex aircraft. Northrop produced and delivered,

In its first flight at Edwards AFB on July 30, 1959 the N-156F exceeded Mach 1 in a shallow dive without the use of afterburner. Lew Nelson, the Northrop test pilot reported, "transition to supersonic, as with the T-38's, is so smooth you have to watch the instruments to know when it happens." (Northrop)

through USAF Foreign Military Sales (FMS) program, approximately 1,100 single seat F-5A and two seat F-5B aircraft to 15 countries over an eight-year period. Lockheed and Northrop, feeling that a need for an F-104 and/or F-5 replacement existed, began funding design studies to create a follow-on aircraft.

Lockheed's CL-1200 design was based on the existing F-104. By retaining much of the proven F-104 design and including existing parts and equipment, the Lockheed replacement was especially attractive to countries already operating the F-104.

Northrop's design was a totally new fighter tailored to the NATO force modernization requirements for the 1970s. Northrop's new airplane was the P530 Cobra. The twin turbofan engine P530 was designed as a multi-purpose air superiority and ground attack fighter, with a top speed of approximately Mach 2. Northrop estimated the total NATO market at approximately 3,000 fighters, with 1,000 of these being susceptible to replacement with a Northrop product. Northrop's plans included a multi-national co-production effort. Northrop proposed the prototyping and development of the P530 be financed entirely by the participating nations. Northrop's plan was for a two phase program. The first phase, pre-production and development, validated the design and flight tested several prototypes. The second phase was production, qualification, and delivery of a combat ready aircraft. Northrop required firm orders for 400 aircraft, or $100 million in development funding before going ahead with the program.

The U.S. Department of Defense (DoD) agreed to help back the P530 program by partially funding its promotion and sales. The rationalization for this was the pressing need for a new fighter in Europe and the feeling that no other U.S. aircraft had the potential to fill this need quickly and effectively. Unfortunately, Northrop was later unable to meet several of the government financial support requirements and the money was used elsewhere.

Area ruling (Coke bottle shape) of the fuselage is an application of the Whitcomb theory of area distribution for reduction of drag. The underside of the N-156F seen here shows the pinching-in of the fuselage at the wing root which enabled the T-38, the N-156F, and the F-5 series of aircraft to easily accelerate from transonic to supersonic flight. (Northrop)

## P530 COBRA

The P530 design, totally funded by Northrop, was the first in the family of the aircraft designs which culminated in the YF-17 and F-18. Although its effects were to be far-reaching, the P530 design never progressed beyond the mock-up stage. Its weight was less than 50% of contemporary first-line, twin-engine fighters with nearly twice the range, maneuverability, and acceleration. The price of the P530 was forecast to be competitive in the international market. The origins of the P530 reach back to 1965 when operation analyses were conducted to define the requirements for an advanced F-5 light weight, multi-role, tactical fighter designated the N-300. The N-300 was very similar in concept to the F-5E, but featured a highmounted wing of F-5 planform with a modest sweep angle. It included small wing leading-edge extensions (LEX) in order to maintain the excellent handling characteristics of the F-5E Tiger II.

The function of the LEX was to set up vortices over the upper wing surfaces to scrub it clean of the sluggish boundary layer air and to exploit the phenomena of vortex lift at the high AOA (Angle of Attack), certain to be encountered during hard maneuvering. The engine inlets were set well forward on a stretched F-5 fuselage, although this in turn required long inlet ducts to the two General Electric G15-J1A1 engines. After a year of testing and analysis, the configuration was changed to a high wing location for maximum ordnance flexibility as well as improved maneuverability.

By 1967 the N-300 design had been given a new designation – P530. Northrop's Manager of Advanced Systems, Walt Fellers, was responsible for seeing the P530 through an emerging series of design refinements. The project office was headed by Lee Begin, who had been active in the N-156 design program. John Patierno led the aero-propulsion effort, and Jerry Huben handled configuration integration.

The LEXs on the P530 were greatly extended, while the inlets were cut back to a position beneath the wing. Wing leading edge fillets also appeared at this time. In March 1968, the proposed engine was changed to the GE15-J1A2, with a thrust of 10,000 pounds. This was an increase of 2,000 pounds over the thrust of the J1A1 model. The aerodynamically curved, wing leading-edge extensions (LEX) and their longitudinal slots had been increased in size for greater lift and airflow control. The single fin and rudder assembly had been replaced by twin tail fins canted outward. At high AOA, a single fin would be completely blanked by the fuselage. Twin fins were an answer to this problem and provided the desired lateral directional stability at high angles of attack. Finally a more advanced avionics system was added.

During 1969 the LEXs were recontoured to increase lift and improve the vortex bursting character, which enhanced the stability and control characteristics at high angles of attack (AOA). The twin fins were enlarged and moved forward on the fuselage to a position where, at high AOA, they would

| ADVANCED TACTICAL FIGHTER EVOLUTION | | | | | LWF PROPOSAL | | |
|---|---|---|---|---|---|---|---|
| HIGH WING FORWARD INLETS | LARGER LE EXTENSION UNDERWING INLETS | TWIN VERTICALS LARGER LE EXTENSION | CONTOURED LE EXT LARGER TAIL | REFINED FUSELAGE SHORTER INLETS | P600 TWIN ENG LWF  P610 SINGLE ENG LWF | YF-17 PROTOTYPE | F-18  LAND-BASED VERSION |
| N-300 | P530 | P530-2 | P530-2 | P530-3 | | | |
| 1966 | 1967 | 1968 | 1969 | 1970 | 1971-72 | 1973 | 1976 |

**This evolution genealogy shows the Northrop designs which led to the YF-17 and later the F-18.**

**Bathed in fluorescent oil, this early model of the P530 illustrated the use of wing leading edge extensions and slots. The efficient performance of the aerodynamic shape of the P530 evolved over more than 5,000 wind tunnel hours. (Northrop)**

not be blanked by the horizontal tail surfaces. The cockpit was moved forward for better pilot visibility and for improved avionics and gun system installations.

In 1970 the P530 was in its fourth design iteration. The engine under consideration was the J1A5 with a thrust of 13,000 pounds. The fixed cone inlet of the earlier models was replaced with a two dimensional fixed ramp inlet. Slots were formed in the LEX next to the fuselage to allow sluggish boundary layer air to be sucked away before it could enter the air intakes. The LEX gave a distinctly "hooded" appearance to a mock-up produced at this stage, giving the P530 its name "Cobra."

The design of the P530 had always been focused on air-to-air combat superiority in order to provide the balance of energy and maneuverability needed to achieve maximum multi-role capability. Since the inception of the program, Northrop engineers had concentrated on the tactical analyses of air-to-air combat and on the identification of those factors which drove design optimization. Thousands of hours of wind tunnel testing verified each step along the way. The P530 program represented a unique concept for a military program. It had been designed to international industrial and economic requirements as well as to military requirements. The Mach 2-class P530 was an ideal offering to foreign buyers from several standpoints; a multi-purpose weapon capability, a low-cost initial outlay, large-scale industrial participation, and advanced technology exchange.

The advanced technology features of the P530 were not only its shape and engines, but also its electronics, armament, and materials. An advanced radar enabled pilots for the first time to detect airborne targets at low, as well as high, altitudes. The design included nine external weapon stations which gave it a wide range of armament carriage for both close ground support and air superiority mission roles. Graphite-epoxy composites were used to lower cost and reduce weight. Northrop drew on its experience in graphite-epoxy composites, both in the laboratory and in actual use on F-5 reinforced structural components, and applied the new materials to the P530 design.

On January 31, 1971, Thomas Jones, then Chairman and President of Northrop, in a letter to the AF Chief of Staff, Gen. John Ryan, offered to build and flight test at company expense two pre-production P530-3 fighters. Jones' letter started by specifically describing the P530 proposal and then noting that as good as its predicted performance was in a heavily loaded condition (i.e., with a full avionics complement), performance could be greatly enhanced by the removal of all systems not essential to a single purpose air-to-air combat fighter variant. The letter concluded by offering two austere P530s to the AF for $15-million if the Dutch (who then were seriously considering the aircraft as a Lockheed F-104 replacement) would make the $100-million commitment required to initiate production. Later, an Air Staff assessment of the Northrop proposal determined it to be roughly similar to Lockheed's, with the exception of Northrop's superior corporate financial position and the required longer time from program go-ahead to first flight (27 months for Northrop v/s 12 months for Lockheed). The situation became somewhat more confusing on March 16, 1971, when Jones again wrote Gen. Ryan offering a further modified P530, known as the P530-4. This aircraft, referred to as an Air Research Vehicle (ARV), was proposed for use as an advanced aerodynamic and propulsion systems testbed. Importantly, initiation of this project no longer was contingent upon a foreign financial commitment.

The AF reacted to the newest Northrop proposal as they had the previous ones, with reservations. On March 24 the company was advised that no positive action would be taken on the P530/ARV because of budgetary constraints, but that it would be reviewed again along with other similar proposals at a later date.

The initial rejection of the P530 was a blow to Northrop, particularly in light of the fact that articles in both the U.S. and European aviation press then were stating (inaccurately) the AF was about to fund the Lockheed CL-1200. Some consolation later was garnered from the fact the AF subsequently elected to fund the General Electric GE-15 (later redesignated YJ101). This afterburning turbojet engine, now funded, would be the engine used in the P530.

By the end of April 1971, it was clear that for either the Northrop or the Lockheed proposals to become serious contenders for funding, they would have to become part of larger prototyping programs. This, in turn, would have to be supported by Congress.

Northrop now made another significant move by proposing a program for a single-engine design, the P-610. This aircraft was based on the original P530, but was considerably smaller and weighed only 17,000 pounds. (some 7,000

The Cobra wooden mockup, completed in December 1972, had the curved wing leading edge extensions which gave the hooded cobra look from which it got its name. The longitudinal slots on either side of the fuselage controlled airflow to the engine intakes. (Northrop)

pounds. less than the P530-4). Although a preliminary design that needed refinement, the P610, with its lighter weight and smaller profile, was expected to have a better maneuverability envelope than its predecessor, particularly at high speeds. While the various aircraft companies maneuvered to generate interest in their endless parade of proposals, the DoD, under direction from Congress and the Secretary of Defense, began to explore the options provided by the old fashioned process of prototyping. This approach, which effectively had been killed during the late 1950s, when hardware costs began to soar apparently out of control, now was taking on renewed importance because of the highly competitive elements of the acquisition process. This, coupled with the lengthier operational service careers that had become more commonplace for aircraft and other military vehicles, dictated that a more detailed and intense review of frontline hardware be made.

Northrop's enthusiasm for the idea of a fighter prototype fly-off program proved fairly guarded, as they were less than enthusiastic about the chances of their P-610 design. Regardless, the company found itself being forced slowly by circumstances beyond its control to significantly alter the P530 from the design that had evolved over the preceding five years. By mid-1971, Northrop was faced with an imminent AF prototype program calling for an austere 20,000 pounds-or-less air superiority fighter with only a questionable chance for follow-on production. Northrop also knew that, if such a program materialized, competition would be intense. In effect, Northrop had to be one of the winners or otherwise lose credibility with the large NATO market. Yet another concern was possible AF preference for a single-engine fighter. The single-engine P610 was, of course, a hurried reaction to this concern, as was a program to pare at least 5,000 pounds off the twin-engine P530's empty weight. Northrop was not comfortable submitting either design for competition.

The instrument panel of the P530 mockup shows off the modern avionics planned for the production version. A Head-Up-Display (HUD) in the upper center of the panel, a radar display below the HUD, a projected map display to the right of the radar, and a radar warning display above the map display were a few of the many aids designed to help the pilot navigate, detect targets, and fire weapons. (Northrop)

Above: Nine ordnance positions (one on each wing tip, three under each wing, and one on the fuselage centerline) gave the P530 a wide flexibility of stores for both air-to-air and air-to-ground missions. (Northrop)

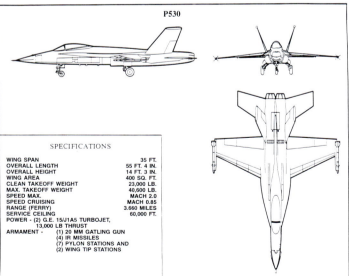

P530

### SPECIFICATIONS

| | |
|---|---|
| WING SPAN | 35 FT. |
| OVERALL LENGTH | 55 FT. 4 IN. |
| OVERALL HEIGHT | 14 FT. 3 IN. |
| WING AREA | 400 SQ. FT. |
| CLEAN TAKEOFF WEIGHT | 23,000 LB. |
| MAX. TAKEOFF WEIGHT | 40,600 LB. |
| SPEED MAX. | MACH 2.0 |
| SPEED CRUISING | MACH 0.85 |
| RANGE (FERRY) | 3,660 MILES |
| SERVICE CEILING | 60,000 FT. |
| POWER - (2) G.E. 15/J1A5 TURBOJET, 13,000 LB THRUST | |
| ARMAMENT - (1) 20 MM GATLING GUN (4) IR MISSILES (7) PYLON STATIONS AND (2) WING TIP STATIONS | |

A PICTORIAL HISTORY • 11

## INTERNATIONAL FIGHTER AIRCRAFT

For some time the DoD had been concerned about the losses, to the French and other countries, of the U.S. competitive position in the international sale of combat aircraft. Because of increasing foreign competition, DoD studies and other efforts were initiated to examine an improved Northrop F-5 (the F-5-21 prototype, later designated F-5E, had flown in 1969) and a less complex and lightened version of the McDonnell F-4E for foreign military sales.

In September 1969, the DoD requested approval from Congress for initiation of development by the Air Force of a new international fighter, appropriate for use by U.S. allies. This request was to fill the need over the following five to six years for approximately 325 aircraft for South Vietnam, South Korea, Taiwan, and other allies. The DoD stated it was in the best U.S. economic, political, and military interests to develop and produce a modern, relatively inexpensive, defensive fighter-interceptor and supply it to U.S. allies under FMS (Foreign Military Sales). This, in the DoD view, would enable the militaries of U.S. allies to become more self-sufficient and less dependent upon the U.S. for their defense.

This aircraft could also be marketed to other friendly countries. The new IFA (International Fighter Aircraft) must have adequate capabilities to handle the existing threat, be inexpensive, and simple to maintain and operate. After several DoD source selection reviews of proposals from Lockheed (design based on F-104), Vought (design based on the F-8), McDonnell Douglas (design based on an austere F-4E), and Northrop (F-5-21, an advanced F-5 design); the Air Force announced on November 20, 1970, that the Northrop candidate, now designated F-5E, had been selected as the new IFA. The F-5E was easily the smallest and least expensive aircraft of those proposed, and also the one that came closest to meeting the restrained IFA mission requirements. In addition, the F-5A and F-5B were still in production, making transition to production of the F-5E a simple matter. The F-5E found a receptive market overseas, not only as a replace-

**The F-5E "Tiger II" was faster and more maneuverable than its older brothers, the F-5A and F-5C. (Northrop)**

ment for the F-5A, but also as a front line fighter in countries which were receiving Northrop aircraft for the first time. Before delivery of the first aircraft, orders for the F-5E exceeded 450. In 1973 F-5E production rate reached 15 aircraft per month; in mid-1975, 18 per month. In 1975, $49 million was funded for development of the two-seat F-5F as a companion fighter-trainer for the E.

Instead of the 325 aircraft envisioned to be produced over a five year period, over 1,419 F-5Es and two seat F-5Fs have been produced for 19 different countries.

## LIGHT WEIGHT FIGHTER AIRCRAFT

In 1970 a new weapon system development policy was formulated by the DoD which required the manufacture of a prototype as an early phase of the design program. This new prototyping policy was designed to reduce risks to the government by demonstrating hardware before it was committed to production. This new "fly before buy" policy, which was first used in the AX program that pitted the Northrop A-9 design against the Fairchild-Republic A-10, was a return to the policy last used in 1958 with the fly-off between the Republic F-105 and the North American F-107. Among the first projects to be selected by the USAF for fly before buy development was a Light Weight Fighter (LWF).

As a result of the experience gained in the F-5 program, the DoD had committed itself to the idea that a quite capable, inexpensive, light weight fighter could be developed to supplement the larger, more expensive FX fighter being developed. The FX fighter program resulted in the development and procurement of the F-15 series of aircraft.

The A-9, Northrop's entry into the AX program, was a conventional design resembling a big T-37, or the later Soviet Su-25 design. (USAF)

The YA-10, Fairchild-Republic's winning AX design, included twin tail mounted engines. (USAF)

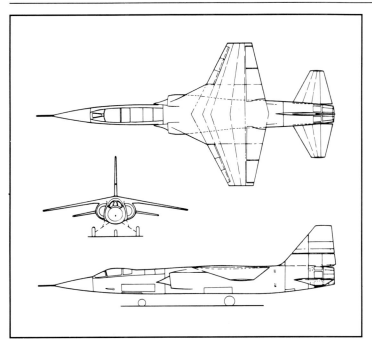

**The three view drawing of Lockheed's CL-1200 shows the design similarities with the F-104. This allowed Lockheed to use many of the F-104's components in the design. (Lockheed-California Co.)**

The FY 1972 Appropriations Bill included $12 million for the initiation of prototype development of both a light weight fighter aircraft and a medium STOL transport aircraft. As a result of this funding, the Air Force officially launched the Light Weight Fighter (LWF) Prototype Program.

A Request for Proposal (RFP) was issued by the Air Force in the fall of 1971. It called for an aircraft highly specialized in the visual, clear weather, day fighter role for air superiority over the battlefield, with particular emphasis on light weight and low cost. The new fighter would have to sustain high rates of turn and increased supersonic maneuvering capabilities, while retaining the ability to accelerate rapidly.

No performance requirements were stated for flight above Mach 1.6. The LWF's role was to be confined to air superiority. The LWF's external stores capability was to be limited, and electronic equipment kept to a bare minimum. The prototype program was to evaluate advanced technology and design concepts, determine aircraft capabilities, and establish potential operational utility.

Although no commitment to production was included in the LWF program contracts, the DoD was tentatively exploring the possibility of a new hi/lo fighter aircraft mix, not only for the U.S. Air Force, but also for the U.S. Navy. No matter how capable a fighter truly is, there is a numerical strength level below which an air force is unable to carry out its assigned tasks. Their chosen future fighter, the large and expensive F-15 Eagle, which had not yet entered USAF service, was quite simply unaffordable in the numbers required. The answer was to compromise with a hi/lo mix: adequate numerical strength containing a significant level of high technology.

The Eagle would provide the hi-tech element, backed by a force of austere but affordable fighters optimized for close combat. This had immediate repercussions on the LWF program. No longer was it to be just a technology demonstration; the winner was to be developed and procured in large numbers by the USAF. In this scenario, the LWF might be considered as a supplement to the Air Force's F-15 and the Navy's F-14.

Five of the nine U.S. aerospace companies receiving RFPs responded. These were Lockheed, Northrop, Boeing, Vought, and General Dynamics. The designs were delivered to the Air Force on February 18, 1972, and a preliminary analysis was completed during the following three weeks. On March 18th the results were delivered.

The Air Force had concluded that the Boeing Model 908/909 design was the number one contender and the General Dynamics Model 401 design was a very close second. Northrop's twin-engine Model P-600 design followed in third place, Vought's Model V-1100 was fourth, and the Lockheed Model CL-1200 design was a distant fifth.

The findings were further analyzed and after some revision, the General Dynamics airplane was the first choice, the Northrop airplane the second choice, and the Boeing airplane

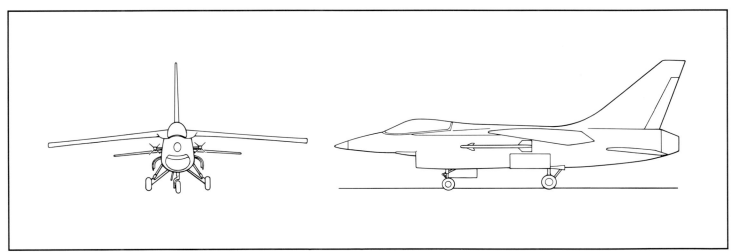

**This drawing of Vought's entry shows its similarity to the successful F-8 aircraft. (Vought Corporation)**

third. This new ranking was based partially on the fact that the General Dynamics and Boeing designs were very similar, and in order to satisfy one of the main objectives of the prototype concept, the decision was made to award the second contract for the Northrop aircraft whose two engine, twin vertical tail design differed from the single engine conventional tail design of both Boeing and General Dynamics. At the conclusion of industry competition and proposal evaluation in April 1972, the Air Force announced Northrop and General Dynamics as winners, each to produce two flying prototypes. General Dynamics was awarded $37,943,000 for two Model 401's. Northrop was awarded $30,978,715 for two Model 600's. The General Dynamics and Northrop LWF designs were now identified as the YF-16 and YF-17, respectively, and soon to be known as Air Combat Fighters, or ACFs.

The test program was based at Edwards AFB with participation by the contractors and the USAF. Determining the winner of the competition between the two pairs of prototypes was to be based on two criteria: performance and cost. Two Joint Test Teams (one for the YF-16 and one for the YF-17), were in place at Edwards AFB during the flight test program. The teams were each made up of members from the contractor, AFFTC (Air Force Flight Test Center), TAC (Tactical Air Command), NASA, and AFTEC (Air Force Test and Evaluation Center). The head of both test teams was the AFFTC Joint Test Force Director. Each Test Team was monitored by a Test Management Council (TMC) consisting of the SPO (System Project Office) Director and senior representatives from TAC, AFFTC, NASA, AFTEC, and the contractors.

The Northrop family of aircraft, on display at Northrop's Hawthorne facility on April 4, 1974, included the YF-17 (foreground), the F-5E in camouflage paint, the F-5B on the extreme left, the T-38 in the rear, and the F-5A behind the YF-17. (Northrop)

A PICTORIAL HISTORY • 15

## YF-16

The General Dynamics Model 401, the company model number of the YF-16 design, was the end product of a lengthy design study and wind tunnel test program. In the design of the YF-16 emphasis was placed on small size and light weight using the application of state of the art technologies resulting in balance of combat capability (aerodynamic performance) and lowest possible mission weight.

The key configuration elements of the YF-16 design were a bottom inlet located under the cockpit, wing-body blending, variable camber wing, and a single vertical tail. In addition, the single-engine configuration was nearly 4,500 pounds lighter than a twin-engine configuration. Following the signing of the official contract on April 13, 1972, construction of the two prototype YF-16's was initiated. Construction of the two prototypes took 21 months; this was an exceptionally short period for a high performance, high technology aircraft. Official taxi-out ceremonies with the No.1 aircraft (72-1567) took place on December 13, 1973, at General Dynamics Fort Worth, Texas production facility (Air Force Plant 4). The two prototype aircraft were disassembled in order to fit in the cargo bay of a C-5A and delivered to Edwards AFB.

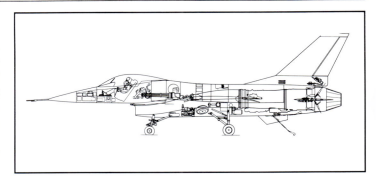

The C-5A carrying the first prototype YF-16 arrived at Edwards AFB on January 8, 1974. It was then reassembled and prepared for initiation of its flight test program. The second YF-16 (72-1568) was cleared for flight test following its delivery to Edwards AFB on February 27, 1974.

Physically, with the exception of their respective paint schemes, the two prototypes were identical. The -1 aircraft was painted a bright red, white, and blue scheme; and the -2 aircraft was painted in a General Dynamics-generated sky camouflage scheme made up of air superiority blue broken up by random areas of off-white, simulating clouds.

**The two General Dynamics model 401 aircraft, given the USAF design number YF-16, were the YF-17's competition in the Light Weight Fighter (LWF) Program fly-off. (USAF via Marty Isham)**

# YF-17

Upon issuance of the RFP, it became evident that the P530 development work was applicable to the LWF. Following a refinement of the P530 design in accordance with the detailed RFP requirements, Northrop submitted its proposal to the Air Force in January 1972. The new design was designated P600. A single engine P-610 was originally submitted, but the proposal was resubmitted as the P530 design following an expressed foreign interest in the original twin-engine P530.

The P600 incorporated the GE15-J1A5 engine (later designated YJ101) with thrust increased to 14,400 pounds. In addition, the wing area was reduced to 350 square feet to improve supersonic performance.

The YF-17's design features optimized to meet the demands for high energy levels, good maneuvering characteristics, excellent stall-post-stall characteristics, and turning capability in the air battle arena, were as follows:

- Maneuvering flaps, large leading edge extensions
- Twin tails for directional stability,
- Cockpit design for increased load factor tolerance

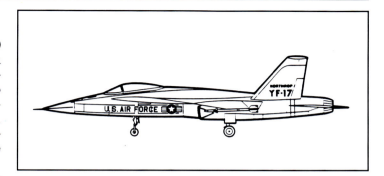

- A wing/engine inlet integrated to allow the engines to perform under the most difficult conditions of low airspeed, high altitude and large yaw rates and high pitch angles.

The basic wing planform was tailored to provide good flying qualities and a high degree of spin resistance. The planform was combined with the large leading edge extension (strake) which gave a remarkable increase in the maximum lift coefficient and reduced buffet intensity, and decreased the drag encountered at supersonic speeds and during high maneuvering situations. The leading edge extension approxi-

The YF-17 was Northrop's entry into the Light Weight Fighter fly-off. (Northrop)

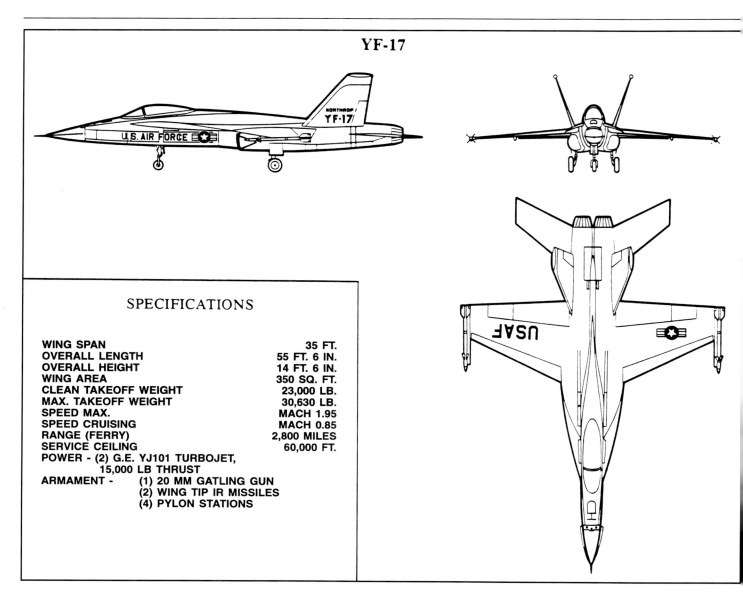

# YF-17

## SPECIFICATIONS

| | |
|---|---|
| WING SPAN | 35 FT. |
| OVERALL LENGTH | 55 FT. 6 IN. |
| OVERALL HEIGHT | 14 FT. 6 IN. |
| WING AREA | 350 SQ. FT. |
| CLEAN TAKEOFF WEIGHT | 23,000 LB. |
| MAX. TAKEOFF WEIGHT | 30,630 LB. |
| SPEED MAX. | MACH 1.95 |
| SPEED CRUISING | MACH 0.85 |
| RANGE (FERRY) | 2,800 MILES |
| SERVICE CEILING | 60,000 FT. |
| POWER - (2) G.E. YJ101 TURBOJET, 15,000 LB THRUST | |
| ARMAMENT - (1) 20 MM GATLING GUN (2) WING TIP IR MISSILES (4) PYLON STATIONS | |

mately doubled the lift coefficient of the basic wing. The maneuvering flaps provided an additional lift above that created by leading edge extension and basic wing planform.

The YF-17 also incorporated a rather unique feature known as differential area ruling. Normally, fuselage area ruling is performed to obtain higher transonic acceleration rates. However, for this aircraft it was felt that differential area ruling was desired for the upper and lower fuselage areas to optimize the supersonic turning capability. The combination of all these design features resulted in improved turn rates.

The twin vertical tails were canted slightly outboard. This feature provided the high degree of directional stability required during the dynamic maneuvers expected from this type of aircraft. With the twin tails, one vertical was expected always to be in the free stream air to provide directional stability, regardless of the angle-of-attack and sideslip.

The YF-17 was powered by two General Electric YJ101 dual-shaft axial flow afterburning turbojet engines rated at approximately 15,000 pounds thrust each. This GE15 design engine eventually entered production (for the F-18) as the F404. A considerable amount of pre-flight phase testing was accomplished in the YF-17 program prior to the actual beginning of inflight testing. The large amplitude flight simulator located at Northrop's Hawthorne, California production plant, the spin and free flight test tunnels at Langley, and the Cal-Span variable stability T-33 aircraft were used to accomplish the testing prior to the in flight phase at Edwards AFB.

Wind tunnel development testing of the P530/YF-17 design extended over a period of eight years and amounted to 9,700 hours total time. Wind tunnel testing was used to develop the hybrid wing, the engine/airframe integration, and the positioning of the vertical and horizontal stabilizers for optimal high angle-of-attack performance. The design allowed the YF-17 to fly what the Northrop test pilots referred to as the "hang and look" maneuver (also called the Cobra maneuver), in which the aircraft flew tail first with a pitch angle 105 degrees. This maneuver was demonstrated at various airshows following the completion of the LWF fly-off. The Su-27 is capable of performance of a similar maneuver.

Both the YF-17 aircraft, No.1 and No.2 seen here, were built in Northrop's Plant 3 in Hawthorne, California. The fuselage was built in three major assemblies with the wings and tail surfaces added. (Northrop)

The aircraft were trucked to Edwards AFB for flight preparation and checkout. No.2 is seen here on July 30, 1974, passing through downtown Los Angeles, with the Los Angeles City Hall in the background. (Northrop)

YF-17 Prototype No.1 (72-1569) rolled out of the Hawthorne plant and was trucked to Edwards AFB for flight test preparation. The initial airborne testing of the YF-17 began with the first flight of the No. 1 airplane on June 9, 1974, at Edwards AFB, with Chief Test Pilot Hank Chouteau in the cockpit. Incidentally this was also the first flight test of the GE YJ101. Following the 61-minute first flight, Chouteau remarked, "When our designers said that in the YF-17 they were going to give the airplane back to the pilot, they meant it. It's a fighter pilot's fighter." Two days later, the YF-17 flew at supersonic speeds in level flight without afterburner – a first for any U.S. built airplane. First flight of Prototype No.2 (75-1570) occurred on August 21, 1974.

Both aircraft were flown simultaneously to evaluate operational factors and subsystems development-evaluation, and to assess reliability-maintainability.

As with the YF-16, the two prototypes were identical except for their paint schemes. The No.1 aircraft was painted with the same aluminum color used on the F-5 series; and the No.2 aircraft was painted in a Northrop sky camouflage scheme made up of an overall off-white color, broken up by areas of medium gray wrapping around the aircraft.

A PICTORIAL HISTORY • 19

A third airframe was constructed for aircraft proof load testing. (Northrop)

Aircraft No.2 is seen here on July 19, 1974, being prepared for painting. (Northrop)

## THE FLY-OFF

The fly-off competition between the YF-17 prototypes and the YF-16 prototypes officially began at the time of the respective first flights of No. 1 aircraft. Comparisons were primarily paper in nature throughout the fly-off, and because of exceptional political circumstances and the undisguised Air Force dislike of the entire lightweight fighter program (which they considered to be a direct and very serious threat to the F-15), the aircraft were never actually flown against each other.

By mid-January 1975, an accelerated seven-month flight test program was completed for the YF-17. A total of 288 flights, or 345 hours of flight test time, had been recorded. Prototype No.1 made 191 flights; Prototype No.2 made 97 flights. The earlier aircraft was used for a majority of the test program elements, including flight controls development, flutter, airframe-propulsion compatibility, armament-propulsion compatibility, performance, and stability and control. The second aircraft was used for structural loads, stall/post stall, and handling qualities during tracking. During the fly-off tests, the airplanes were flown against other operational Air Force fighters and used in a variety of weapons tests including bomb delivery, air-to-air gunnery, and missile launch.

In similar tests completed in mid-December 1974, the two YF-16s logged a total of 330 missions, amounting to 417 hours of flight time including 13 hours and 15 minutes at supersonic speeds, reaching altitudes in excess of 60,000 feet, achieving maneuvering g forces of 9, and accelerating on several occasions to speeds in excess of Mach 2.

By the time the fly-off was over, a relatively large number of pilots had flown both aircraft types. Company and Air Force test pilots had logged the majority of hours, with Navy and miscellaneous government pilots logging the rest. Key inputs from these professionals played a major role in the final deci-

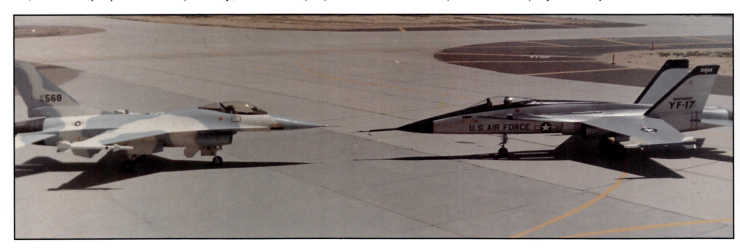

Like the YF-16s, the two YF-17 prototypes each had their own distinctive paint scheme. The No.2 aircraft (72-1570) aircraft was painted in the "Northrop Sky" camouflage made up of an off-white color broken up by areas of medium gray. The No.1 aircraft (72-1569) was painted with the same aluminum color used on the F-5 series aircraft. (Northrop)

sion making process. Adversary aircraft flown against the YF-16s and YF-17s included the Cessna A-37B, the McDonnell F-4E, the Convair F-106, the MiG-17, and the MiG-21.

After the last mission of the fly-off, the analysis of gathered data and pilot input was begun. On January 13, 1975, Secretary of the Air Force John McLucas made the announcement – the General Dynamics YF-16 had been picked as the light weight fighter (Air Combat Fighter) competition winner based on the program's two primary criteria, performance and cost.

The information obtained during the flight test program clearly established the high-performance and low-cost characteristics of the YF-17. The Air Force decided that the YF-16 was better suited for its specific operational requirements, notably the supplementing of its air-to-air F-15 fighter force and including considerations of its force structure mix and logistics commonality with other aircraft already in service (i.e. the same basic engine as the F-15). With this decision,

all Northrop's work seemed to have been in vain, as it now appeared likely that General Dynamics would pick up the export orders which Northrop had been so confidently expecting for the Cobra.

The YF-17 performed creditably, demonstrating a Vmax of Mach 1.95, a combat ceiling of over 50,000 feet (15,250m), and a peak load factor of 9.4g. Handling was all that could be desired. There were no adverse departure tendencies, and an AOA of 68 degrees was attained in maneuvering flight, which incidentally was far beyond what the opposing YF-16 could reach.

Northrop also felt it had operated at a distinct disadvantage due to the fact that the YF-17 had entered the flight test program four months later than the YF-16, thus forcing the company to compress all testing into seven months rather than ten. NATO was awaiting the final lightweight fighter competition ruling before initiating the final stage of its decision making process, and it had requested that the Air Force make

With No.1 in the background, No.2 is being prepared for flight. The screens over the engine intakes allowed ground crew to safely work on the aircraft and perform checks with the engines running. (Northrop)

a choice no later than December 1974. Another Northrop disadvantage was that the YF-17 was equipped with a prototype engine (YJ101), whereas the YF-16's F100 was a proven powerplant with significant logged time.

However, all was not lost. The United States Navy (USN) was in a similar straitened predicament over the cost of the F-14 Tomcat, and at the same time was faced with replacing large numbers of both F-4 Phantom II multi-role fighters and Vought A-7 Corsair II attack aircraft. What the USN really wanted was a new single design for both roles but, for budgetary reasons, Congress dictated the Navy examine derivatives of both LWF contenders. The Navy acquiesced on condition that both manufacturers teamed with companies experienced in the design of naval aircraft. General Dynamics linked with Ling Temco Vought, while Northrop joined forces with McDonnell Douglas. The Northrop project was numbered P630.

The YF-17, which had been designed to a concept of mission versatility providing a high degree of air-to-ground as well as air-to-air capability, went on to win the Navy air combat fighter competition in which multi-mission suitability was a greater consideration. The YF-17 design was selected by the Navy to be modified for carrier operations and was redesignated the F/A-18. The F/A-18 replaced the F-4 and A-7 in the U.S. Navy and Marine Corps.

The YF-17 had wing tip missile rails, like the F-5, which allowed them to carry AIM-9E Sidewinder missiles on each wing tip. (Northrop)

## YF-17 FLIGHT TEST EXPERIENCE

Predicted flight loads and measured flight loads were very close. As a result of the loads correlation, the pilot was given 100 percent design operating load factor capability within the air combat maneuvering envelope.

Flight test proved that the YF-17 was free from all aeroelastic and aeroservoelastic instabilities with and without wingtip missiles. Reduced root stiffness was experienced in the horizontal tail, which resulted in varying degrees of reduced roll rate at high dynamic pressures, especially at supersonic speeds below 30,000 feet.

No major structural failures or discrepancies occurred during the flight test program of either No. 1 or No. 2 aircraft.

One failure involved the loss of a portion of the leading edge flap-to-wing seal strip on both left and right wings. This occurred during the first flight on the number one aircraft and was due to insufficient stiffness of the thin aluminum strip. Stiffening doublers were added to the seals, and no further trouble was encountered.

During flight number nine on aircraft No. 2, the canopy glass separated from the frame and left the airplane. It was determined that the attach bolts for the glass had been improperly installed.

During flight number 187 on aircraft No.1 the entire canopy left the airplane. The cause of failure was inconclusive as the canopy was not recovered for inspection; it appeared that the canopy actuator failed in tension. The canopy/latch was modified, and a cockpit switch to interrupt the power supply to the actuator was installed to prevent an actuator malfunction.

During flight number 74 on aircraft No.1, a piece of the graphite-composite at the base of the leading edge of the left vertical stabilizer broke off due to defective bond. The damaged area was repaired by trimming and cleaning the damaged area and wrapping fiberglass over foam blocks to form the aerodynamic shape. All repairs were accomplished, and periodic inspections of the repaired area indicated no further problems.

This shot of No.1 refueling behind a KC-135A shows the comparative size of the YF-17. Indicator lights on the head-up-display frame tell the pilot when the boom is latched, ready, or disconnected. (Northrop)

The fiberglass antenna pods on the tip of the vertical stabilizer deteriorated, and the vertical tip rib cracked due to higher than anticipated dynamic loads. Redesign and replacement was necessary.

During flight test the YJ101 engine/YF-17 installation demonstrated surge-free operation throughout a wide range of angle of attack (AOA) and sideslip angles imposed during engine/airframe compatibility testing and the stall/post stall test phase.

During the Stall/Post Stall test phase, AOAs from -14 to +63 degrees and angles of sideslip up to 39 degrees at an AOA of 50 degrees were encountered without compressor stall.

The YJ101 engine/YF-17 installation demonstrated excellent windmilling start capability throughout flight test.

Surge-free transient operation during both static ground and high altitude flight conditions were accomplished. Numerous throttle commands (idle to maximum, chop to idle, burst to maximum before engine stabilizes) were performed without incident at the functional-check-flight condition of 0.9 Mach at 45,000 feet. Successful throttle bursts and chops were accomplished as high as 50,000 feet and as fast as 1.5 Mach during level flight conditions.

Flight tests show that the fuel system performed satisfactorily. The fuel feed system fulfilled all demands of engine flow rates and supply pressures throughout the operating envelope, in level and maneuvering flight attitudes. Satisfactory engine operation at 30,000 feet on gravity feed was also demonstrated. The transfer system met design objectives. The vent system required some modification to prevent it from venting fuel overboard.

Aerial refueling experience was excellent and unanimously judged to be the easiest of any aircraft previously flown by the members of the Joint Test Team. The task was enhanced by the forward boom receptacle location which allows the pilot better visibility to more easily place the aircraft in the proper envelope for boom contact. Holding position was effortless due to the excellent power response and flying qualities of the YF-17.

During flight test all control system functions were satisfactory. Some refinements were incorporated to remove mi-

The lowspeed characteristics of the YF-17 were quite good, even with two 600 gallon fuel tanks, as evidenced by this bottom view of No.1 being refueled by a KC-97. (Northrop)

nor deficiencies uncovered during various phases of the flight test program. The automatic flight control system demonstrated system/airframe compatibility in ground tests and during flight after minor modifications in structural filters. Adequate gain margin exists between the gains being flown and limit cycle divergence points.

Initially, roll control performance was judged inadequate at transonic speed at low altitude. This was due to aero-elastic influences on the aileron and rolling tail effectiveness. Improved roll performance was demonstrated following removal of Mach number scheduled limits from the ailerons and an increase in rolling tail authority. A change in lateral control forces was made to optimize tracking performance and resulted in fully satisfactory longitudinal/lateral control harmony.

Yaw control system characteristics were improved as a result of changes in augmentation feedback and roll-to-yaw crossfeed gains, and were satisfactory. Further gain optimization was desirable to achieve fully satisfactory tracking characteristics during high roll rates.

Flap control system characteristics were satisfactory. Normal acceleration feedback was added to the maneuvering mode command schedule to minimize interaction with

The YF-17 accomplished weapon release testing of many weapons. Seen here No.1 releases a MK84 low drag general purpose bomb from the left wing inboard pylon. (Northrop)

aircraft short period pitch characteristics. In response to pilot requests, momentary override functions were added to (1) enhance unloading and acceleration from a high AOA (angle of attack) condition, (2) prevent flap motion during gun firing, and (3) increase aerodynamic braking ability following touchdown.

Ramp diverter door actuation and throttle control system characteristics were satisfactory.

The accelerated flight test program precluded proper integration and evaluation of the fire control system. The optical sight system experienced a series of power supply failures during initial use which rendered it inoperative during much of the flight test program. Additionally, the instantaneous vertical field-of-view (IVFOV) was considered inadequate by the pilots. The power supply failure was resolved near the conclusion of the test program. However, insufficient time was available for performance evaluation. A dual combining glass was implemented which corrected the IVFOV deficiency. Sufficient operational time was obtained to demonstrate tracking during gunfire without pipper blur and jitter. An F-5E type Lead Computing Optical Sight (LCOS) was installed with a fixed reticle for tracking purposes. A second LCOS system was installed with fixed flight parameter inputs and dual combining glass to qualitatively evaluate aircraft/fire control system interface compatibility. System performance was satisfactory.

The fire control radar performance was not evaluated due to the accelerated flight test program and the aforementioned problem with the optical sight system. An operational optical sight system was required to evaluate the radar performance. The radar, however, was observed to be operational during the several opportunities afforded.

No.1 YF-17 performs the "Hang and Look" maneuver (later called the Cobra maneuver). In this maneuver, later made famous at air shows by the Russian Sukhoi Su-27, the YF-17 hangs in space with little forward or upward velocity. (Northrop)

**No.1 in its aluminum paint seen here north of Edwards AFB with AIM-9Es on the wing tips. (Northrop)**

During flight test the M61 Gun System performed properly and without malfunction for both ground and in-flight firings.

Engine/gun compatibility was demonstrated in that no smoke ingestion at the engine inlet was observed.

Pilot comments indicate that the cockpit environment was acceptable during gun firing, and that no flash obstruction occurred. Some visual obscuration due to gun gas occurred during gun firing at air targets. This gun gas and the resulting residue accumulation on the windscreen could be eliminated by redesign of the gun gas diffuser.

Successful carriage and jettison of the MK-84 LDGP store was demonstrated from the inboard pylon on the YF-17 airplane. Good separation characteristics were demonstrated with minimal airplane response following store release.

AIM-9E missile compatibility was successfully demonstrated regarding loading and firing. Missiles were launched with good separation, causing negligible aircraft reaction and no adverse engine effects.

The Digital Air Data Computer (DADC) and angle-of-attack (AOA) system performance was considered to be outstanding. No re-calibration was required for static source correction. AOA data was reliable and accurate. Several random failures of the DADC were apparently induced by high temperature stress resulting from high environmental equipment bay temperatures experienced during ground operation when cooling was not available.

As a result of flight test, Joint Task Force pilots report physical tolerance to load factor was increased by 2 g's with the leg/seat arrangement of the YF-17 over conventional seating arrangements. Rudder pedal forward adjustment proved insufficient and was increased by 1.5 inches. During long duration flights pilots experienced fatigue in the upper legs due to inadequate thigh support. This was solved by individual seat cushion wedges and was a result of the inability to adjust the rudder pedals forward sufficiently. Tape engine instruments did not satisfy the test pilots because of difficulty in detecting rate-of-change, which they were accustomed to with conventional dial type instruments. The cockpit, however, was considered basically satisfactory.

These views show large wing Leading Edge Extensions (LEXs), which spawned the Cobra nickname, and the aft positioning of the twin engines. (Northrop)

The two YF-17s are seen here in this shot from a tanker boom pod. (Northrop)

On January 15, 1974, the No.1 aircraft visited the 57th Fighter Weapons Wing at Nellis AFB. (Don Logan)

While at Nellis, the aircraft was parked at the end of the Thunderbird ramp, next to the Thunderbird T-38s. (Don Logan)

**Right:** No.1 over the desert north of Edwards AFB. A majority of the Light Weight Fighter fly-off took place in the Edwards AFB range complex over the California desert. (Northrop)

**Below:** This view of the two YF-17 contrasts the No.1 silver aircraft with the No.2 camouflaged aircraft. (Northrop)

No.2 is seen here over the Borax mine north of Edwards. The spherical object visible between the vertical stabilizers covers the spin recover parachute. (Northrop)

The No.2 aircraft photographed on display at the Edwards AFB Open House and Air Show in November 1975. (Don Logan)

Sun shades on movable stands were used to protect the test crew from the hot Edwards sun. (Mick Roth)

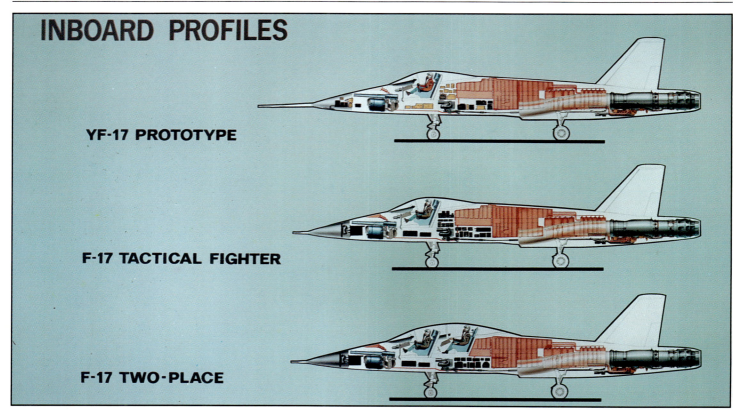

Had the YF-17 been selected as the fly-off winner, both a single seat tactical fighter and a two seat fighter/trainer version would have been built. (Northrop)

Though a two seat version was never built, the Cobra mockup seen here on October 17, 1974, was modified to the two seat configuration. (Northrop)

## NASA AND THE F-17

Following the creation of NASA, The Dryden Flight Research Center (DFRC) at Edwards AFB was involved in programs using various U.S. military aircraft: Lockheed F-104A Starfighter, McDonnell F-4 Phantom II, General Dynamics F-111A, Lockheed T-33 Shooting Star, Northrop F-5A Freedom Fighter, Vought F-8C Crusader, McDonnell Douglas F-15A Eagle, Northrop YF-17 Cobra, and the McDonnell Douglas F-18A Hornet. NASA had other programs that were military related," such as the Blackbirds (using the YF-12s and SR-71s), XB-70A, and the TACT F-111. DFRC also used airplanes from abandoned projects, such as the Northrop A-9A.

DFRC flew numerous brief programs using military airplanes. During 1970 and 1971, the DFRC undertook a study of high-lift flaps as aids to transonic maneuverability with a series of tests on F-104, Northrop F-5A Freedom Fighter, and Vought F-8C Crusader aircraft. Wind tunnel tests simply were not reliable for this purpose, and the data resulting from flight test was useful in the design of new military aircraft. DFRC's work in this area led to the definition, by the DoD, of "agility" criteria for fighter turn rate, buffet, maximum lift, and handling qualities. This criteria was used in the development of the new generation of fighter aircraft including the McDonnell Douglas F-15A Eagle, the General Dynamics F-16A, the Northrop YF-17 Cobra, and the McDonnell Douglas F-18 Hornet.

After sitting briefly in storage DFRC flew the first YF-17 during 1976, at the request of the Navy, for base drag studies and to evaluate the maneuvering capability and limitations of the aircraft. NASA pilots, each flying the plane at least once, and engineers examined the YF-17's buffet, stability and control, handling qualities, and acceleration characteristics.

The YF-17 surprised many of the DFRC's pilots trained on earlier combat aircraft. "I was astounded," one center pilot recalled. "That airplane really is a generation ahead of anything else. It's got twice the performance of current-day airplanes like the F-4, and some of the others. It'll climb twice as fast, and it'll burn half the fuel-just phenomenal."

The No.1 aircraft was flown by NASA during 1976. The aircraft is seen here after receiving NASA markings. (NASA – Robert L. Lawson Collection)

## F-18 PROTOTYPE

During the early days of the LWF program, it became apparent there was demand for a similar light weight fighter aircraft to fulfill certain Navy requirements. The requirement for a Navy LWF, like the AF LWF, had been brought about by the expected high cost of the new Navy fighter being developed under the VFX (fixed-wing fighter experimental) competition. The VFX program design was for another expensive, large, and very complex fighter. This program came about as the result of a Navy decision to avoid acquisition of the Grumman built F-111B, a navalized version of the Air Force TFX (General Dynamics F-111). The VFX program ended with the procurement of the Grumman F-14, at that time, the most expensive and complex production fighter in history.

The VFX program highlighted the fact the Navy had two serious problems: (1) the new fighter, at over $30-million per copy, was quickly using up the Navy acquisition budget; and (2) there was a strong move in Congress for an increased number of combat aircraft. In response to the latter, various House and Senate members had begun openly to advocate a mix of both VFAX (fixed-wing fighter experimental) and Navy LWF aircraft in order to keep up with the fast growing Soviet Union's military arsenal.

With the interest in an increased number of aircraft for the Navy, the service had to look beyond the expensive F-14 and accept a fighter that was less complex, cheaper, and maybe less capable.

During the summer of 1973, James Schlesinger became the new Secretary of Defense and mandated that both the Air Force and the Navy study LWF design. In August 1973, the Navy was instructed by Congress to pursue a lower cost alternative to the F-14. The following month, the Navy was instructed to request proposals from the U.S. aerospace industry.

The Navy presented a proposal to the Congress and Schlesinger calling for a "stripped" version of the F-14 to be called the F-14X, in an attempt to protect the F-14 from funding reductions. This aircraft deleted the AIM-54 Phoenix air-to-air missile and its associated fire control system, but re-

The similarities between the YF-17 (72-1570 seen here in F-18 prototype markings) and the F-5E can be seen in this photo, taken in 1977 during Navy testing. (Northrop)

Visible here on 72-1570 in a down (deployed) position are the full leading edge maneuvering flaps. (Don Logan)

tained the AIM-7 Sparrow radar-guided AAM system. In the Navy's opinion, it was the best and cheapest solution to supply the increased number of aircraft.

Congress and Schlesinger disagreed with the Navy's solution. On May 10, 1974, the House Armed Services Committee deleted the Navy's request for $34-million to start the new VFAX program which called for a totally new fighter to complement the F-14, and instructed the Navy consider one of the two AF lightweight fighters (YF-16 or YF-17) undergoing the fly-off flight testing at Edwards AFB. The Navy still pressed for the VFAX program; but after Senate Joint Committee on Appropriations released a strong statement in support of the Navy LWF buy, the Navy gave in.

The Navy found itself in the position of having to procure whichever LWF design the AF liked the best. This put the Navy in a position similar that of the TFX (F-111) program, where it was forced to accept a navalized version of an Air Force design. While the Navy explored its few remaining options and debated the Congressional instructions internally, Northrop reached a partnership agreement with the McDonnell Douglas Corporation for a navalized version of Northrop's YF-17. A Northrop internal information announcement, dated October 7, 1974, said in part:

"Northrop Corporation and McDonnell Douglas Corporation today announced that the two aerospace firms have entered into an agreement under which they will jointly develop and propose an air combat fighter for the U.S. Navy which is based on the YF-17 design ... Under the teaming agreement, McDonnell Douglas will have prime contract responsibility for a carrier-suitable version of the YF-17 to meet the requirement of the proposed NACF (Naval Air Combat Fighter). Northrop will have prime contract and design responsibility for YF-17 variants for use by NATO nations and other allies."

"Over the past 20 years, both companies have concentrated their fighter design efforts on the development of high-performance, twin-engined aircraft. Both companies also have extensive experience in the international marketplace. Northrop has produced and delivered more than 2,200 twin-engine tactical and trainer aircraft, which are in service with the USAF and more than 20 nations around the world. The company's latest combat aircraft are the F-5E International Fighter and its two-place companion, the F-5F."

"McDonnell Douglas has produced more than 4,500 of the F-4 Phantom fighter, including the F-4B model, which has been in operational service with the U.S. Navy since 1961. Its newest combat aircraft is the F-15 air superiority fighter for the U.S. Air Force."

Similarly, General Dynamics and Vought joined in an agreement that would give Vought the right to produce a navalized version of the YF-16.

The Navy, almost from the very beginning, had taken a strong stand in favor of the YF-17. The Navy liked the Northrop design because of its redundancy with two engines, room for growth and expansion, and its mission adaptability. Additionally, the choice by Northrop of McDonnell Douglas, a successful producer of Navy fighters, as partner reassured the Navy that navalized YF-17 design could fulfill Navy requirements. During the Air Force ACF fly-off, Navy pilots flew both the YF-16 and YF-17 prototypes and concluded the YF-17 was the preferable design for Navy aircraft carrier requirements.

When the YF-16 was announced as the final LWF competition winner, the response by the Navy was universally negative. Informally, the Navy's position was that it had no intention of buying the navalized version of the F-16 as set forth by Congress and Schlesinger, and the Navy immediately started a campaign either to kill the Navy LWF program or to force Congress to allow the service to acquire the navalized YF-17.

During the four months following the AF award to General Dynamics, the Navy negotiated with Congress and the Secretary of Defense. On May 2, 1975, a compromise agreement called for the Navy to develop a navalized derivative of the Northrop YF-17 (known originally as the P-630 and later

**As part of the F-18 development program, 72-1569 flew with TOP GUN at NAS Miramar during 1977. (Northrop)**

as the McDonnell Douglas Model 267). This aircraft was designated originally as the F-18 Navy Air Combat Fighter (NACF) and later as the dual mission F/A-18 Navy/Marine Corps-strike fighter. The initial plan provided for distinctive F-18 fighter and A-18 attack versions, but later, all hardware differences were eliminated with the fighter and attack variants differing only in the armaments carried for specific missions.

Congress gradually accepted the Navy's position that technical changes required to make the YF-16 carrier suitable canceled out the commonality between AF and Navy fighters desired by Congress. The Navy convinced Congress this would be less cost effective than the development of navalized F-17 design. During 1976, full scale development contracts were awarded by the Navy to McDonnell Douglas for airframe construction and final assembly (with Northrop as the major subcontractor, responsible for construction of the main fuselage, which made up approximately 40% of the airframe) and to General Electric for the F404 engine.

**During 1976, while No.1 (72-1569) was flying with NASA, No.2 was painted white and used by Northrop as a sales demonstrator in their attempt to sell the aircraft world wide. (Northrop)**

The twin vertical stabilizers on the YF-17 were canted outboard in order to remain effective at very high angles of attack (AOAs). (Northrop)

Above and below: No.2 YF-17 (72-1570) was seen at numerous air shows during 1976. (D.J. Fisher)

This view shows 72-1570 painted in the same blue and white scheme later applied to the first test F-18 aircraft. (Northrop –Robert L. Lawson Collection)

Seen on rotation, the wing LEXs resembling a cobra's hood, show the basis of the aircraft's nickname – COBRA. (Northrop – Robert L. Lawson Collection)

On October 9, 1976, No.2 was in England at the Farnborough Air Show. (Michael A. France)

**Above:** Air to ground ordnance including laser guided bombs, general purpose bombs, and Maverick missiles was loaded on the pylons on the right wing. (Michael A. France)

**Left:** The left wing carried a 600 gallon fuel tank, AIM-7 Sparrow, and an AIM-9 Sidewinder. (Michael A. France)

A PICTORIAL HISTORY • 37

**In early 1977, No.1 (72-1569) transferred from NASA to the U.S. Navy and flew with TOP GUN at NAS Miramar as part of the F-18 Hornet development program. (Roy Lock)**

**No.2 seen here painted in a camouflage scheme similar to the scheme on TOP GUN's F-5Es with NAVY on the right side. (Roy Lock)**

**MARINES was painted on the left side of the aircraft. (Roy Lock)**

Right: During 1977, 72-1570's tail carried the F-18 Hornet team markings. (Don Logan)

Below: While flying as part of the F-18 development program, the YF-17 carried the Northrop name and the Cobra insignia. (Don Logan)

72-1570 flew at the NAS Miramar Air Show on October 28, 1977. (Don Logan)

A PICTORIAL HISTORY • 39

600 gallon auxiliary fuel tanks could be carried on the inboard wing pylons. (Northrop)

72-1569 carried VX-4's XF tail code and F/A-18 PROTOTYPE markings while flying out of Point Mugu NAS in late 1978. (Don Logan Collection)

72-1569 was retired to the Western Museum of Flight near Northrop's facility in Hawthorne, California. (Craig Kaston)

Seen here on May 14, 1978, 72-1570 painted light gray while flying out of NAS Lemoore, California. (Don Logan)

72-1570 was marked A-18 PROTOTYPE during its stay at NAS Lemoore. It was later retired to the National Museum of Naval Aviation at NAS Pensacola, Florida. (Don Logan)

A PICTORIAL HISTORY • 41

## F-18 HORNET

The F-18 was intended to replace both Navy and Marine Corps F-4 Phantoms for the primary missions of fighter escort and interdiction. The F-18 was also intended for eventual use, during the mid-1980s, in the attack role as replacement for the A-7 Corsair II. A two-place fighter-trainer version was also programmed to be built. Cost estimates were predicated on a potential purchase of 800 aircraft.

Northrop's decision not to continue as the prime contractor (giving MCAIR 60% of the production responsibilities while Northrop retained 40%) in the program resulted from Northrop's realization of their lack of experience in building carrier suitable aircraft. Additionally, major programs more in line with Northrop's production abilities were looming on the horizon, and the company did not have the physical plant to handle the production capacities that eventually would be demanded for several programs. Northrop retained the right to be prime contractor for the export version of the F/A-18 which was to have been designated F-18L, with MCAIR being the primary subcontractor with the split being 60% Northrop, 40% MCAIR.

The Navy had committed itself to the McDonnell Douglas/Northrop aircraft, with the production responsibilities split between the two companies. Northrop would build the vertical tail assemblies and the center and aft fuselage subassemblies, and MCAIR would build all remaining components including the wings, while serving also as the final assembler. The General Electric advanced YJ101 turbojet engine, designated F404, was designated to be used in the F/A-18. This engine represented the state-of-the-art in conventional turbojet engine design, had significantly more thrust than its predecessor, and was fully capable of providing the power required by the new Navy fighter.

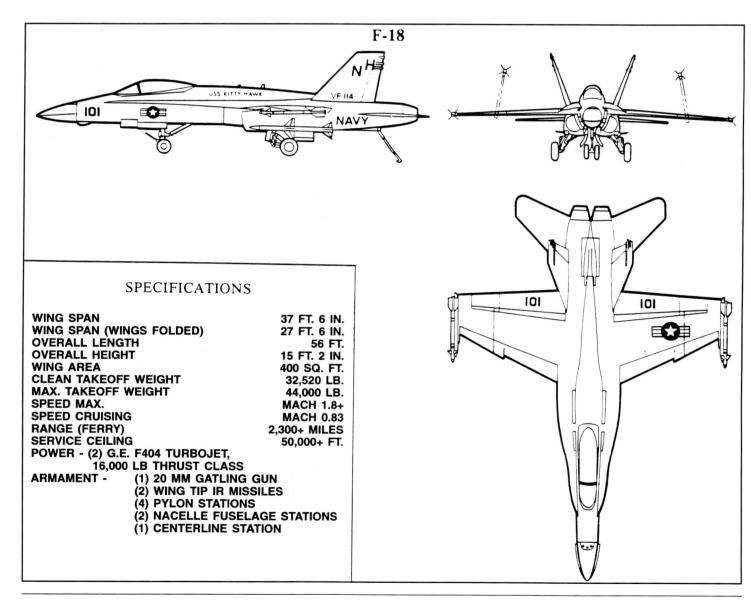

### SPECIFICATIONS

| | |
|---|---|
| WING SPAN | 37 FT. 6 IN. |
| WING SPAN (WINGS FOLDED) | 27 FT. 6 IN. |
| OVERALL LENGTH | 56 FT. |
| OVERALL HEIGHT | 15 FT. 2 IN. |
| WING AREA | 400 SQ. FT. |
| CLEAN TAKEOFF WEIGHT | 32,520 LB. |
| MAX. TAKEOFF WEIGHT | 44,000 LB. |
| SPEED MAX. | MACH 1.8+ |
| SPEED CRUISING | MACH 0.83 |
| RANGE (FERRY) | 2,300+ MILES |
| SERVICE CEILING | 50,000+ FT. |
| POWER - | (2) G.E. F404 TURBOJET, 16,000 LB THRUST CLASS |
| ARMAMENT - | (1) 20 MM GATLING GUN |
| | (2) WING TIP IR MISSILES |
| | (4) PYLON STATIONS |
| | (2) NACELLE FUSELAGE STATIONS |
| | (1) CENTERLINE STATION |

**This view of a VFA-105 F/A-18 shows how similar the F/A-18 design was to the design of the YF-17. (Norris Graser)**

**The F/A-18 continued with the large leading edge extensions (LEX) and the twin outward tilting tail. (Mick Roth)**

**Though only mockups of the two seat F-17 were made, a two seat version of the F-18 was produced as both the F-18B and F-18D. (Norris Graser)**

**72-1570 flew in Canadian markings while Northrop was trying to sell F-18Ls to Canada. (Northrop)**

The new, navalized YF-17 was redesignated F-18 (later, F/A-18 once the two role/two version program was combined). This was done in an attempt to represent the aircraft as a totally new design unrelated to the Air Force YF-17. In fact, the F-18 was a new aircraft. It retained the basic configuration of the YF-17, but had few components related to those found on the YF-17. Major changes for the F/A-18 included a wider fuselage which made room for the addition of a Hughes manufactured multi-mode radar, provisions for the AIM-7 Sparrow air-to-air missile and its associated fire control system, 4,460 lbs. of additional fuel (giving the aircraft a total of 10,680 lbs. of fuel in four fuselage tanks and two inboard wing section tanks), an articulated two wheel nose gear incorporating a towbar to be used for catapult launches from the aircraft carrier, a carrier capable arresting hook, folding outer wing panels, increased wing area (2.5 ft. longer which increased the wing area by 50 sq. ft.) to accommodate the increased aircraft gross weight, increased structural strengthening to accommodate the requirements of carrier operations, and a completely redesigned and significantly strengthened carrier capable landing gear. These changes added over 10,000 lbs. to the YF-17's 23,000 lb. gross weight. Other changes were incorporated to improve the aircraft's performance at the low speed and increase maneuverability. The wing LEXs were redesigned and increased in total area. The programmed deployment angles of the leading and trailing edge flaps which were increased from 30 to 45 degrees and the ailerons were re-programmed to droop at a maximum angle of 45 degrees in low speed flight. The stabilators which were increased in area, but were revised with a lower aspect ratio. A dog tooth was added to the leading edges of both the wings and the stabilators to cause vortex generation. The resulting design of the F/A-18 compared quite favorably to the Navy's original VFX requirement.

With certain exceptions, principally in avionics equipment provisions, the F-18 was similar to the YF-17 aerodynamically and structurally. High-strength aluminum alloys were the primary material, supplemented by extensive applications of graphite composites. A fly-by-wire flight controls system was provided with mechanical backup in the horizontal tail pitch and roll control system.

In the Spring of 1976, plans were formalized with the U.S. Navy and the Department of Defense for a land based version of the F/A-18. It was especially developed by Northrop to satisfy worldwide requirements for an advanced aircraft, operating efficiently from conventional runways on land, with both the performance versatility and economic features of the Navy F/A-18. The land based version, the F-18L, incorporated the same design philosophy applied to the internationally successful F-5 series of fighters. Technological innovations permitted it to be a multipurpose successor to aircraft such as the F-4 capable of performing as an interceptor, as a long-range interdiction aircraft, and in the close-support and air-to-air combat roles.

A key design requirement of the new fighter was the achievement of reliability much higher than the aircraft that it replaced. As a result of designed-in high reliability, reduced maintenance, and reduced manpower requirement followed. Built-in test capabilities of rapid fault isolation and location, and quick ground-level access were among its desirable maintenance features.

Northrop's land-based version was over 6,000 pounds lighter than the Navy and Marine F/A-18. Of this total, 3500 pounds were as a result of the reduced fuel load alone. Elimination of the carrier operations requirement generated other "weight savers." These included a lightweight landing gear, the use of a non-folding wing, a lighter arresting gear, and a simplified avionics package. The reduced weight, coupled with the use of the same twin engines (General Electric F404, 16,000-pound thrust class) and the same aerodynamic lines, contributed to the aircraft's very high performance capability.

The most obvious external difference between the Navy F/A-18 and the proposed land-based F-18L version was the elimination of the wing slots needed for carrier landing approaches at high angles of attack. Both the land fighter and the Navy F/A-18 benefitted from their 60 percent commonality of parts by weight and 85-90 percent commonality in high-use systems. A wide range of weapons options were available on the land based aircraft for both air and ground missions, with a total external stores capacity exceeding 13,000 pounds. Options included AIM-7 Sparrow and AIM-9 Sidewinder missiles. No customers for the F-18L design were ever found. MCAIR, however, attempted to sell the F/A-18 on the international market. By 1984, lawsuits resulting from these attempts by MCAIR to sell their version of the Hornet to foreign customers were initiated by Northrop. The prime question in these lawsuits: "Was the F/A-18 a modified Northrop design or was it a MCAIR design?" The lawsuits were reconciled in May 1985 when McDonnell Douglas agreed to pay Northrop $50 million in return for the rights to be prime contractor on all export versions of the F/A-18. This allowed the two companies to successfully accommodate production obligations for the F/A-18.

The mockup, still in the two seat configuration and with F/A-18L painted on the side, was displayed with a complete load on air to ground ordnance. (Northrop)

# AIRCRAFT DESCRIPTION

**P530 COBRA**

Though never progressing beyond the mockup stage, the P530 design called for construction primarily of riveted aluminum alloy in a semi-monocoque fuselage. Steel and titanium were used in high temperature areas and for areas with a high structural loads. Graphite-epoxy laminates and aluminum honeycomb were also used for certain doors and secondary structures.

Four bladder-type fuel tanks were located in the fuselage, behind the cockpit, and were filled using a single-point pressure refueling receptacle. To minimize the possibility of fuel vapor explosions, polyurethane reticulated foam was used inside all fuel tanks.

The thin, two piece dry wing was of multi-spar construction with thick, machined aluminum skins. The left and right wings were attached to the fuselage with shear bolts. Aluminum honeycomb was applied extensively in panels, leading edge and trailing edge flaps, and ailerons.

The all-movable horizontal tail was an aluminum bonded assembly with honeycomb core. The twin vertical stabilizers were of thick aluminum skin with multiple spars, and the rudders were aluminum-bonded assemblies with honeycomb cores.

Many of the P530 systems were very similar to those proven in the F-5 tactical fighter, using a number of redundant systems. Dual hydraulic systems were provided for the primary flight controls, including a variable-camber high-lift subsystem for maneuvering, takeoff, and landing. An electronic Control Augmentation System (CAS) was incorporated in parallel with the mechanical system. It added or subtracted control travel and rate, as determined by the airplane electronic model program contained in the CAS. This program was designed to produce desirable handling qualities over the entire flight envelope, taking into account airplane loading, speed, and altitude.

As in the F-5, the twin turbojet engines were mounted side-by-side in the aft section of the fuselage. The P530 en-

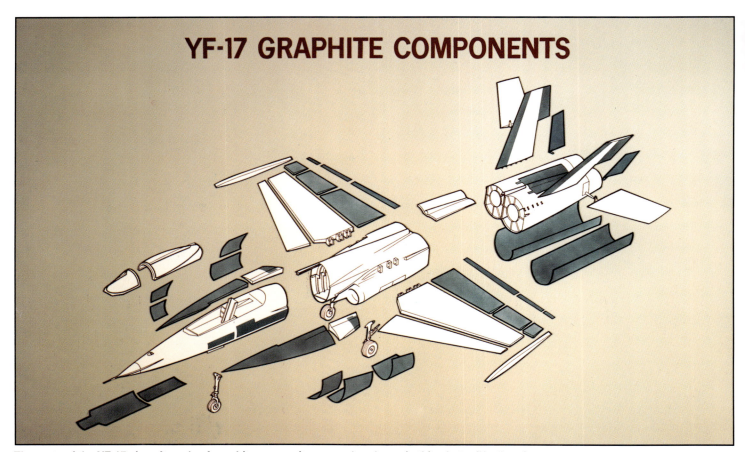

**The parts of the YF-17 aircraft made of graphite composites are colored gray in this photo. (Northrop)**

gines were removed by lowering them vertically, without removing the horizontal tail or boattail as required in the F-5/T-38 design. A three-point mounting arrangement permitted quick engine removal or installation. Titanium firewalls isolated the engines from one another and from the forward airframe compartments. An integrated engine starter and accessory drive system was mounted on the airframe forward of the engines. Each engine could be removed without disturbing the electrical, hydraulic, or gas turbine starter units by disconnecting the engine drive shaft from the power package.

The P530 was to be equipped with a single 20mm Gatling-type gun mounted in the upper forward fuselage. Alternate arrangements included two 30mm cannon or two 20mm cannon. Four missile launcher rails were installed for IR (AIM-9) missile carriage: one on each wingtip and two on wing pylons. A total of seven external pylon stations supplemented the two wingtip stations to provide for carriage of a variety of ordnance or fuel stores.

## YF-17

The YF-17 was a single-place, twin-engine air superiority prototype lightweight fighter aircraft in the 20,000-pound weight class featuring a high thrust-to-weight ratio and low wing loading for maximum maneuverability. Advanced propulsion technology, in the form of the General Electric YJ101-GE-100 engines, was closely integrated with advances in aerodynamics, controls, and structures. The YF-17 was built by the Northrop Corporation, Aircraft Division in Hawthorne, California.

The YF-17 was designed to facilitate servicing and maintenance. Wing, fuselage, and pylon heights provided convenient access from the ground for most of the systems and components. The aircraft could be serviced without the use of maintenance stands. Turn-around and servicing time was reduced by locating servicing points to minimize interference among the ground crews. For example, the single point refueling receptacle was located on the right side of the aircraft allowing the refueling vehicle to be positioned aft of the right wing and away from the gun servicing area.

Scheduled and unscheduled maintenance actions were facilitated by ground level access to all components other than the rudder, vertical tip antennas, speed brake, and the engine top mounts which required the use of workstands for access.

The bottom of the fuselage was 42 inches above the ground, and the lower surface of the wing averaged 63 inches above ground level. The lower surfaces of the wind pylons averaged approximately 50 inches above the ground, providing ease of loading.

Structurally, the YF-17 was manufactured using a riveted semi-monocoque stressed skin design, employing high-strength aluminum alloys as its primary material. Steel and titanium were used in space-limited areas of high temperature or areas of high loading. The fuselage consisted of stressed skin supported by frames, longerons, and bulkheads. Graphite-epoxy composite material was used for engine bay doors, landing gear doors, and various equipment access doors and panels. The forward fuselage equipment section housed the ranging radar, air data computer, battery, and emergency power unit.

The wing consisted of right and left panels attached to the fuselage with shear bolts. The thin dry wing (the wing contained no integral wing fuel tanks) was of thick skin, multi-spar construction, with fuselage attach ribs of welded and machined titanium. Various sections of the wing featured aluminum or Nomex (nylon/phenolic) honeycomb core with graphite-epoxy face sheets. These included sections of the LEX wing trailing edge, trailing-edge flaps, and ailerons.

The all-movable right and left horizontal stabilizers were all aluminum, bonded assemblies with full-depth honeycomb cores, and machine tapered and sculptured skins having an aluminum integral leading-edge wedge, and a full-span machined aluminum channel spar. The vertical stabilizers were of thick skin, multi-spar construction with leading edges and rudders of aluminum honeycomb core, laminated graphite-epoxy spar, and graphite-epoxy face sheets.

Propulsion thrust was supplied by two YJ101-GE-100 afterburner equipped turbojet engines producing approximately 15,000 pounds of thrust each. Increased thrust performance, stability, and reliability were enhanced by the location of the engine air inlets beneath the wing, diversion of fuselage boundary layer air upstream of the air intake through wing root slots, and the inlets fixed geometry design. The two engines were installed in the aft fuselage utilizing two thrust mounts and an aft steady rest for each engine. The engines were removed by lowering them vertically from the engine bay with no requirement for disconnecting any of the empennage control system. The "engine removal" access door and panels were attached with quick-release structural fasteners.

A key feature of airframe/inlet integration was a longitudinal slot through the wing root to allow fuselage boundary layer diversion above the wing as well as below the fuselage. The inlet installation with slot allows a narrow boundary layer gutter, resulting in a low drag installation. The side fuselage mounted ramp prevents interaction of fuselage boundary layer with the inlet shocks.

A ramp bleed system was used to eliminate the adverse effects of shock/ramp boundary layer interaction at supersonic speeds. Inlet cowl angles minimized spillage drag, and the internal lip shape was designed to prevent flow separation at high angle of attack operation.

The short subsonic diffuser duct with a gradual transition to the engine face provided high pressure recovery. Bay purging air (one percent of duct air) was removed from the duct ahead of the compressor face.

Ingestion of runway debris was prevented by a vortex inhibitor consisting of two jets of compressor bleed air. The jet nozzles were located in the fuselage forward of the inlets. The jets eliminated ground vortices by modifying the flow field below the inlet. Throughout the YF-17 flight test program there were no incidents of engine ingestion of runway objects.

The accessory drive system consisted of two airframe-mounted, engine-driven gearboxes interconnected by a common starter gearbox. During engine removal, the gearboxes remained in the aircraft while the drive shaft was mechanically disconnected to avoid disturbing aircraft systems.

Mechanical power was supplied by two independent 3,000-psi hydraulic systems. The left engine drove one system which provided half the power for the flight controls and the leading-edge flap actuators, as well as all of the power for the tricycle landing gear, brakes, gun, and pitch control augmentation system. The right engine drove the second system which provided half the power for the flight controls and the leading edge flap actuators as well as all the power for the trailing edge flaps, nose wheel steering, and emergency main gear extension. An emergency power unit provided flight control hydraulic power in the event of dual engine failure.

Internal fuel was contained in four bladder-type cells located in the fuselage behind the cockpit. A separate system for each tank gave a crossfeed capability. The fuel system had a single point receptacle ground refueling. Air refueling capability was provided through a receptacle concealed behind a retractable door located in the nose, forward of the windshield.

The primary flight controls consisted of ailerons, rudders, and the all-movable horizontal tail. Basic roll control derived from "fly-by wire" ailerons, supplemented by the differential motion of the horizontal tail. All primary control surfaces were powered by dual hydraulic cylinders for redundancy and integrated with an electronic control augmentation system (CAS). Programmed responses from sensors within the system through a digital air data computer (DADC) provided relatively constant stick force per-G throughout the flight envelope. A feature of the control system was the rolling horizontal tail, providing differential deflection of the tail surfaces which assisted in rolling this aircraft around the approximate gun lead angle. Maneuvering flaps, programmed into the DADC, were mechanically interconnected with the horizontal tail to minimize the requirement for manual trim changes.

Flying qualities throughout the flight envelope were enhanced during high lift flight conditions and in transonic regions by the sweptback cambered wing leading edge extensions (LEX), the variable camber wing (the wing camber was changed using automatic, computer controlled, positioning leading and trailing edge flaps), and twin vertical tails canted outboard and positioned forward of the horizontal tail. The

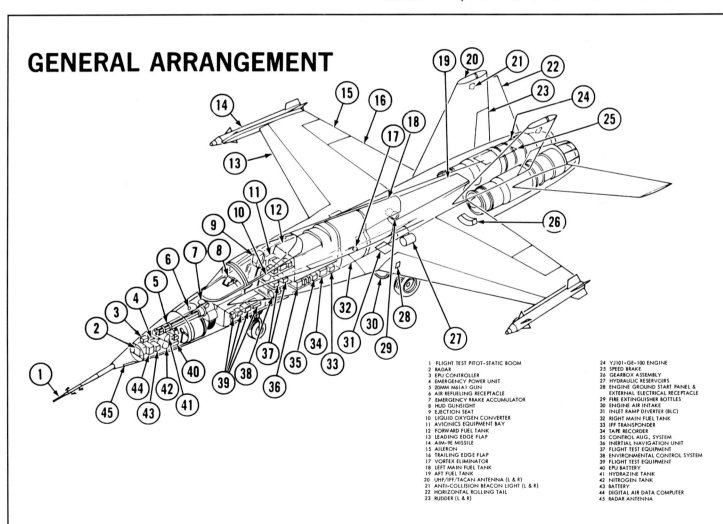

## GENERAL ARRANGEMENT

1. FLIGHT TEST PITOT-STATIC BOOM
2. RADAR
3. EPU CONTROLLER
4. EMERGENCY POWER UNIT
5. 20MM M61A1 GUN
6. AIR REFUELING RECEPTACLE
7. EMERGENCY BRAKE ACCUMULATOR
8. HUD GUNSIGHT
9. EJECTION SEAT
10. LIQUID OXYGEN CONVERTER
11. AVIONICS EQUIPMENT BAY
12. FORWARD FUEL TANK
13. LEADING EDGE FLAP
14. AIM-9E MISSILE
15. AILERON
16. TRAILING EDGE FLAP
17. VORTEX ELIMINATOR
18. LEFT MAIN FUEL TANK
19. AFT FUEL TANK
20. UHF/IFF/TACAN ANTENNA (L & R)
21. ANTI-COLLISION BEACON LIGHT (L & R)
22. HORIZONTAL ROLLING TAIL
23. RUDDER (L & R)
24. YJ101-GE-100 ENGINE
25. SPEED BRAKE
26. GEARBOX ASSEMBLY
27. HYDRAULIC RESERVOIRS
28. ENGINE GROUND START PANEL & EXTERNAL ELECTRICAL RECEPTACLE
29. FIRE EXTINGUISHER BOTTLES
30. ENGINE AIR INTAKE
31. INLET RAMP DIVERTER (BLC)
32. RIGHT MAIN FUEL TANK
33. IFF TRANSPONDER
34. TAPE RECORDER
35. CONTROL AUG. SYSTEM
36. INERTIAL NAVIGATION UNIT
37. FLIGHT TEST EQUIPMENT
38. ENVIRONMENTAL CONTROL SYSTEM
39. FLIGHT TEST EQUIPMENT
40. EPU BATTERY
41. HYDRAZINE TANK
42. NITROGEN TANK
43. BATTERY
44. DIGITAL AIR DATA COMPUTER
45. RADAR ANTENNA

"hybrid" midwing was moderately sweptback, with a slight negative dihedral, and situated well aft on the fuselage. The wing trailing edge was fitted with conventional ailerons and landing flaps. The sweptback horizontal tail was positioned below and aft of the wing, with a negative dihedral, and could be deflected differentially to complement roll control in flight. A speed brake was located on the upper aft fuselage between the twin vertical stabilizers.

Individual left and right throttles could be connected to a master throttle and operated together with the single master throttle lever. The cockpit was equipped with a zero-zero rocket catapult ejection seat. The seat back was positioned to improve rearward visibility with the back angle improving pilot high-G tolerance. The pilot was provided with 360 degrees visibility at and above eye level; 13 degrees over fuselage nose, and 55 degrees over the side.

Basic armament consists of an M61A1 20mm Gatling gun pallet mounted on the centerline of the aircraft in the forward nose section and AIM-9 series missiles carried on wingtip launcher rails. Two pylons under each wing provided carriage of external fuel tanks and/or weapons. The centerline station capability, which was included in the production design, was not installed in the two prototypes. A 600-gallon external fuel tank could be installed on each inboard wing pylon to extend combat range. A computing sight provided a heads-up display (HUD) of target steering and attack.

## DIMENSIONS

| | |
|---|---|
| Wingspan (w/missiles) | 37.7 ft |
| Wingspan (w/o missiles) | 35.0 ft |
| Horizontal Tail Span | 22.2 ft |
| Length (overall) | 62.7 ft |
| Height (overall) | 14.5 ft |
| Wheelbase | 17.2 ft |
| Wheel Tread | 6.8 ft |
| Tail Down Angle (Takeoff and Landing) | 12.0 degrees |

## WEIGHTS

| | |
|---|---|
| Empty | 17,180 lb |
| Basic (empty plus unusable fuel) | 17,290 lb |
| Operating Empty (includes: crew, unusable fuel, oxygen, engine oil) | 17,560 lb |
| Basic Mission Takeoff (includes: full internal fuel, ammo, missiles) | 24,580 lb |
| Ferry Mission Takeoff (includes: full internal fuel, full 600 gal inboard pylon tanks, w/o ammo and missiles) | 32,710 lb |
| Maximum Takeoff (includes: full internal fuel, full 600 gal inboard pylon tanks, ammo, missiles) | 34,280 lb |

## Engines

The YF-17 was powered by two YJ101-GE-100 low bypass twin-spool, continuous bleed, afterburner equipped axial flow turbojet engines each rated at approximately 15,000 pounds static thrust (un-installed). The engines were specifically designed to provide high thrust characteristics in the transonic region while maintaining N1 (low pressure) rotor speed to minimize drag and improve fuel consumption during combat maneuvering. Components were front frame containing the variable inlet guide vanes, 3-stage N1 (low pressure) compressor, mid-frame, 7-stage N2 (high pressure) compressor utilizing three stages of variable stators, annular-type combustor incorporating a carbureted fuel supply for smoke free operation, single-stage N1 (low pressure) turbine, single stage N2 (high pressure) turbine, and the afterburner section. Continuous bleed air from the N1 compressor allowed the engine to be internally self-cooled; 4th stage N2 compressor air was used to assist cooling of the N1 turbine. Hydraulically actuated variable exhaust nozzles automatically provided proper nozzle opening from intermediate thrust through all ranges of afterburner operation.

## Engine Air Induction System

The air inlet induction system for each engine consisted of an oblique shock inlet and subsonic diffuser duct. The inlets were side mounted under the wing leading edge extension (LEX) and were designed to provide reduced inlet Mach number during high supersonic speeds and best performance during maneuver and cruise. A two-dimension fixed geometry inlet with a 75 degree vertical ramp provided for maximum performance in the transonic flight regime. A longitudinal slot through the wing root allowed fuselage boundary layer air diversion above the wing as well as below the fuselage. A ramp bleed system for each inlet eliminated adverse effects of shock/ramp boundary layer interaction at supersonic speeds. A DADC controlled air exit door was provided on the top surface of the LEX. When in the closed position, a 55 degree opening was provided; open position was 155 degrees except with DADC failure which would fully open the door to 205 degrees. If DADC failure occurred at subsonic speeds, the effect of a full open air exit door on engine performance was negligible. In normal operation the door started to open at approximately 1.4 Mach. The subsonic diffuser duct provided high pressure recovery. Approximately 1% of duct air was used for engine bay purging.

## Engine Nozzle Flap

The engine was supplied with an exhaust nozzle external flap arrangement with each flap supported by the engine only at the flap trailing edge. The flap's function was to provide a fairing from the aircraft lines to the nozzle exit. The leading edges of the nozzle external flaps were hinged to a ring which was attached to the airframe. The ring was attached with quick acting latches to facilitate removal. Relative motion between the engine and airframe was accommodated by free angular deflections of the individual flaps.

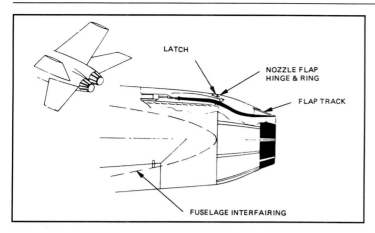
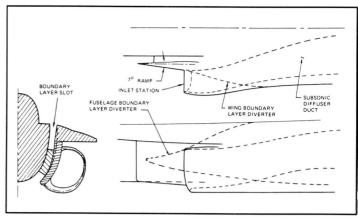

The airframe/nozzle interface was unique in that the hinge ring of the external nozzle flaps was rigidly mounted to the airframe instead of the engine, and the trailing edge of the flaps was conventionally guided in tracks attached to the internal nozzle. This arrangement provided a smooth sealed airframe/nozzle connection to minimize nozzle drag and also allows the engine to expand and flex relative to the airframe. The fuselage between the engines was faired at the flap hinge ring in order to direct the external flow into the region between the nozzles to eliminate internozzle drag.

### Engine Oil System
Each engine has an independent, self-contained oil supply and lubrication system. A sump vented internally within the system (center-venting) prolonged carbon seal life, minimized oil leakage through seals, and maintained positive pressure to make the system insensitive to altitude. Full capacity of the tank was eight quarts of MIL-L-7808G oil; five usable and three unusable. Oil was measured by a dipstick on the fill tube cap and was accessible through a door on each side of the fuselage.

### Engine Bay Ventilation
In flight the purging flow enters the bay from the inlet duct and exits through louvers in the aft fuselage upper and lower surfaces. The effective entrance and exhaust areas were estimated to be four square inches and thirty-two square inches, respectively.

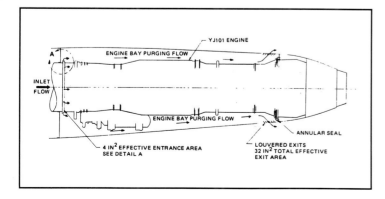

### Engine Control System
The engine control was an electro-hydraulic mechanical type which provided automatic engine operation as determined by throttle positioning. Power settings below INTERMED position (military power) were determined by the main fuel control regulation of the N2 (high pressure) rotor speed. At INTERMED position and through all afterburner (AUGMENT) position ranges, engine speeds were controlled by the N1 (low pressure) rotor speed and EGT. The system was designed to regulate main and afterburner fuel flows, schedule exhaust nozzle area and low/high pressure variable geometry, and monitor N1/N2 pressure compressor rotor speeds and turbine discharge temperature. These functions were performed in response to signals of engine inlet air temperature, compressor discharge temperature, N1 (low pressure) turbine discharge temperature, and feedback signals from each of the controlled functions. The systems automatically prevented the engine from exceeding any limits within the operating envelope. In the event of electrical failure, the system was mechanically controlled to ensure satisfactory operation of the engine from IDLE cutoff to slightly below INTERMED power. The engine was protected from any single control failure to prevent excessive overspeed or excessive overtemperature.

### Throttles
Engine power control was regulated by the throttles, one for each engine, and a master throttle which serves as a single throttle for slaved operation of both the left and right engines. The master throttle grip contained switches for the speed brake, microphone, maneuvering flaps, missile uncage, radar reject, and one flight test function. The left and right throttles could be engaged or disengaged to the master throttle at any position; engagement being accomplished by down and forward motion, disengagement by down and aft motion. If desired, the master throttle could be stowed forward, when disengaged. When either throttle was disengaged and pulled back to IDLE position, it contacted a stop to prevent going into OFF position. The throttle could be locked at idle cutoff. To unlock, down pressure was required before the throttle could be moved forward. A detent located at the INTERMED position must be passed for afterburner (AUGMENT) operation. Afterburner thrust augmentation was provided from MIN

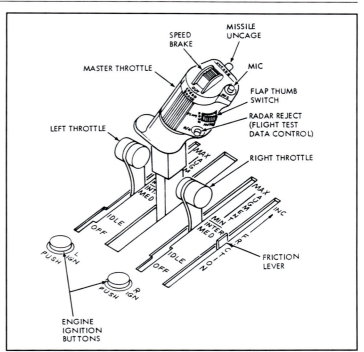

This view of an engine change occurring on 72-1569 shows the simplified access to engine compartment, and the engine trailer which also served as the lift used to install the engine. This was an improvement from Northrop's standard design on the F-5 and T-38 which used a removable fuselage boattail to access the engine for removal. (Northrop)

to MAX positions. Throttle friction could be manually adjusted by the pilot.

### Starter Gearbox
The starter gearbox was driven by a single air-turbine starter mounted on the gearbox. Power output was connected to each engine accessory gearbox through a clutch assembly. Activation of an engine start switch, located in an access panel on the left engine air intake duct, would open a solenoid-operated valve to drive the respective engine clutch and, in turn, drive the engine accessory drive to motor the engine.

### Accessory Drive Gearboxes
The accessory drive gearboxes were independent accessory power systems. Power during start was transmitted to the engine through an interconnecting drive shaft. The hydraulic pump, oil/fuel pump, and generator, all mounted on the gearbox, would rotate during engine start. After engine start the starter and clutch cuts off and the engine drives the gearbox accessories through the interconnecting drive shaft.

### Vortex Eliminator System
Ingestion of runway debris (FOD) was prevented by a vortex eliminator system. Two jet nozzles installed in the centerline lower fuselage forward of the engine air intake ducts aimed a 40-directional flow of high pressure engine compressor bleed air to spoil induced ground vortices by modifying the airflow field below the inlet opening. Operation was automatic with engines running on the ground; however, the nozzles were shut off with the engines at IDLE and when the main landing gear was raised.

### Fire Warning And Extinguishing System
The fire warning and extinguish system provided engine and accessory gearbox compartment fire detection and extinguishing capability. Two pneumatic heat sensing elements attached to each engine (forward and aft) and one element mounted in each gearbox compartment detected a fire. A signal from any element illuminated fire warning lights (red) in respective FIRE PULL handles on the instrument panel. Pulling the handle closed the fuel shutoff valve to the engine and armed the extinguishing system. Two fire extinguisher agent bottles provided a "two-shot" capability to the compartments on one side or "one-shot" for each side. When the FIRE PULL handle was pulled, the agent discharge switch under each handle was accessible. AGENT 1 position under either handle fired the left bottle; AGENT 2 fired the right bottle. If a fire in one engine and accessory compartment was not extinguished by using the first bottle, then the second bottle could be used to extinguish a fire in the same compartment; however, once both bottles were expended, no fire protection was available. Momentary positioning of an agent discharge switch to either bottle discharged the entire contents.

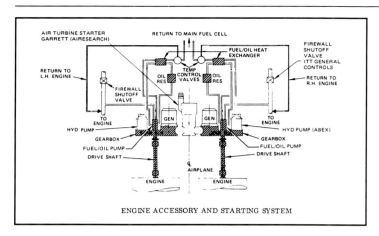

ENGINE ACCESSORY AND STARTING SYSTEM

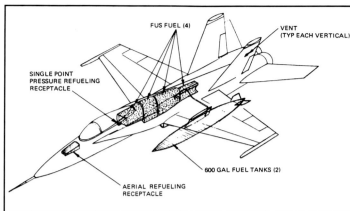

## ENGINE START OPERATION

All engine indicators and start relays were powered by the 28-VDC primary bus. Either engine could be started first; however, starting the left engine was recommended to supply hydraulic power for wheel brakes.

### Ground Start

Starting either engine required external low-pressure air source to drive the air turbine starter gearbox. With the battery switch on, ground starting sequence was initiated by the ground crew (upon signal from the pilot). The engine ignition buttons on the throttle quadrants were not used during ground start. However, the lights individually illuminated (white) whenever the ignition circuit was automatically activated and went out when ignition was completed. When the first indication of high pressure rotor speed was indicated in the N2 tachometer, the throttle was advanced into IDLE position and engine ignition was automatically turned on. Fuel was injected and the exhaust nozzle remained fully open. At approximately 30% N2 the first indication of low pressure rotor speed appeared on the N1 tachometer. The ignition would cut off at about 40% N2. At approximately 52% N2 the starter would cut-off and the engine accelerated to idle rpm. Fuel flow reduced when idle was reached at approximately 70% N2.

### Inflight Start

With throttles at IDLE or above, in flight ignition was automatic and continuous when N2 rpm was between 10% and 40%, and was indicated by the ignition buttons being illuminated. If N2 rpm was above 40%, the respective engine ignition/timer button the throttle quadrant could provide 30 second ignition cycle for engine airstart. When pushed, the button would illuminate to indicate activation. At the end of the 30 second cycle the light went out, and the ignition terminated. If required, the button could be pushed to continue ignition. The buttons could be activated while the engines were running to provide the capability of automatic engine restart during high speed combat maneuvering if flameout occurred.

## AFTERBURNER OPERATION

Advancing the throttle into AUGMENT position turned on afterburner ignition, opened the engine nozzle slightly above an intermediate setting, temporarily lowered the low pressure turbine discharge temperature, and initiated minimum afterburner fuel flow until lightoff occurred. The engine ignition button light would illuminate to indicate ignition. A lightoff detector sensed lightoff and scheduled fuel flow as determined by throttle position. Normal low-pressure compressor rotor speed (N1) and turbine discharge temperature values were established and the exhaust nozzle modulated to hold the turbine discharge temperature limit. On rapid throttle advance from below INTERMED or MAX power settings, lightoff was initiated when high-pressure compressor rotor speed (N2) exceeded 85% and was held until the exhaust nozzle area decreased below a specified value and lightoff was detected. For throttle retard, the exhaust nozzle closed with throttle demand to hold constant turbine discharge temperature, and augmented fuel flow decreased toward the minimum limit. The afterburner fuel pump shut off when fuel flow dropped below a fixed level.

### Fuel System

Approximately 6150 lb of usable internal fuel was carried in four fuselage bladder type tanks. Two external tanks (600 gal each) could be carried, one on each inboard wing pylon. JP-4 fuel was supplied to each engine from a respective main tank and feed system consisting of an inverted flight compartment, boost pump, strainer, and shutoff valve. Crossfeed from either the left or right main tanks was accomplished by use of the crossfeed switch on the left console. Shutoff valve operation was normally controlled by the throttle(s). In an emergency, the valve could be closed by the fire pull handle(s) on the instrument panel. If both boost pumps were inoperative, sufficient fuel would gravity feed to maintain both engines at intermediate power from sea level to 25,000 ft; flame out could occur above 6000 ft during augmented thrust operation. Each engine was fed from a separate main tank, which contained a system consisting of an inverted flight compartment, pump, filter, and shutoff valve. The main tanks were adjacent and connected by a transfer tube. The transfer tube was located at the 900-lb level in each main tank. The length of the transfer tube prevented excessive fuel from transferring between tanks in steep attitudes.

An inverted flight compartment was located in each main tank. One-way flapper valves in the compartments trapped fuel so that the double-ended boost pump could supply fuel to the engines under any flight condition – including inverted flight.

A scavenge pump in each main tank kept the inverted flight compartment full during the majority of flight conditions. The left main scavenge pump, located in the aft part of the tank, was especially valuable in making all of the fuel usable when fuel was low and the aircraft was in a nose-up position. The fuel filters were also mounted in the inverted flight compartments. The filters prevented particles larger than 74 microns from entering the engines. The filter elements were wire mesh type and could be replaced without first draining the fuel from the tanks. A bypass valve and bypass indicator were provided in the filters.

The boost pumps, one for each engine, were AC electric and were protected against overheating by thermal protectors. A pressure switch connected to each pump actuates a warning light on the pilot's caution panel when a pump was inoperative.

Fuel transfer between internal tanks was automatic during normal operation of the system. During automatic transfer, the forward tank transferred first into the main tanks. After the forward tank emptied, the aft tank transferred fuel after a 700 lb fuel level drop in the main tanks. The left and right main tanks were interconnected by a transfer tube at the 1800 lb fuel remaining level (900 lb each tank). Final fuel feed was from the individual main tanks. Transfer of fuel from both wing pylon fuel tanks to the aft tank was accomplished by a transfer pump in each external tank. If the forward transfer pump failed (no caution light), automatic gravity feed to the right main tank occurred as the fuel level in main tanks decreased. If the aft transfer pump failed, a caution light on the pedestal illuminated. Gravity feed to the left main tank was delayed until the fuel level in main tanks dropped below the interconnect transfer tube. To assist gravity transfer of aft tank fuel, it was necessary to select crossfeed from the left main tank and reduce power to obtain a total fuel flow rate of 6000 pounds per hour or less.

## Ground Refueling

Ground refueling of internal fuel was accomplished through a single-point pressure refueling receptacle located in the right fuselage below and slightly aft of the cockpit. The refueling time was about 4 minutes with 50 psi nozzle pressure. External tanks used filler caps for fueling. The aircraft could be defueled through a fitting in the left engine feed line. The internal (and external) fuel transfer pumps and booster pumps, together with the operation of the cross-feed valve, were required for defueling.

## Air Refueling

Air refueling of the internal fuel system was provided through a receptacle located on the nose, forward of the windshield. External tanks were not capable of being air refueled. Refueling time was about four minutes to load 90 percent of the total internal fuel capacity.

The gun bay purge system was automatically activated whenever the receptacle door was opened to vent any fumes or fuel spillage from the gun compartment. An ejector pump in the system scavenged fuel from the air refuel line into the right main tank. The pump was operated by a small amount of fuel bled from the right system fuel boost pump.

## Electrical System

Two 20/24 KVA, 115/200 VAC, 3-phase, 320-500 Hz oil-cooled generators, driven by the accessory drive gearbox of each engine, were used to supply AC power to respective left and right systems. One generator was capable of automatically assuming both AC system power loads if the other generator failed or was turned OFF. External AC power for ground operation was supplied through an external receptacle. DC power was supplied by two 50-amp, 26-32 VDC transformer-rectifiers (TR) and a 24-VDC, 6 amp-hr rechargeable battery. Each respective TR converted AC power to DC power; if one TR failed, the other would continue to supply all essential DC power requirements.

## Emergency Power Supply

Emergency electrical power was supplied by an additional 24-VDC, 3 amp-hr, non-rechargeable (in aircraft) battery. This battery provided a direct source of DC power for operation of the hydrazine emergency power unit (EPU).

## Aircraft Lighting System

Exterior lighting consisted of only the rotating beacon light installed in each vertical tail. Interior lighting consisted of only two cockpit floodlights, mounted aft in the cockpit, and the engine group and hydraulic system indicators. The beacon light switch and cockpit lighting switch were located on the right console.

## Environmental Control System

The environmental control system (ECS) provided cockpit air conditioning and pressurization, avionics equipment compartment cooling, canopy and windshield defogging, canopy seal pressurization, gun breech and gun compartment purging, anti-G suit pressurization, and emergency air ventilation of the cockpit. Fourth stage engine compressor bleed air from each engine was used to supply all conditioned air requirements; however, one engine was capable of operating the complete system. An external bleed air connection permitted prolonged ground cooling operation of avionics equipment without the engines operating. Controls on the right console were provided for operation of cockpit pressurization, temperature, and defog air distribution. A torso outlet on each side of the cockpit provided directional flow of conditioned air and could be adjusted from shutoff to full flow by twisting the knurled outlet fitting. Cockpit pressure altitude was indicated by the cabin altimeter on the pedestal. The cockpit was unpressurized up to 8000 feet. Cockpit pressure was held equal to 8000 feet at altitudes up to 23,500 feet. Above 23,500 feet the system maintained a 5 psig differential. A pressure safety valve in the system automatically protected the cockpit from high and low pressures.

## Hydraulic Systems

Hydraulic power was supplied by two independent left and right systems. Constant pressure variable delivery piston pumps, driven by the respective left and right engine accessory gearbox drive assembly, provided each system with 3000 psi operating pressure. A pressurized reservoir supplied hydraulic fluid to each pump (2.5-gal. left) (1.25-gal. right). Both systems normally operated at all times while both engines were in operation. Loss of either engine, either pump, or either fluid system would not cause loss of the powered flight controls; although a reduction of flight control effectiveness would occur.

## Emergency Power Unit

The hydrazine-fueled emergency power unit (EPU) in the fuselage nose was a self-contained system which operated a variable output (9 gallons per minute) hydraulic pump to provide hydraulic pressure to the right hydraulic system for flight control operation should both generators fail or engines flame-out in flight. The EPU was capable of instant start and would provide approximately 7 minutes of total operation including multiple restarts.

## Emergency Hydraulic Systems

Emergency hydraulic power to boost main landing gear extension was provided through a transfer cylinder containing stored left system hydraulic fluid. The cylinder was powered by right hydraulic system pressure or, if in operation, the EPU. Transfer cylinder activation occurred when the alternate landing gear extension procedure was used.

A separate emergency brake accumulator with stored pressurized left hydraulic system fluid provided alternate wheel braking power and was activated during emergency wheel brake procedure.

## Landing Gear System

The landing gear system provided normal extension and retraction of gear, alternate extension, normal and emergency brake operation, and nosewheel steering. Retraction and extension of gear was accomplished through left hydraulic system power controlled electrically by the landing gear lever on the instrument panel. Normal braking power was supplied by the left hydraulic system and controlled by the brake pedals. Nosewheel steering was hydraulically powered by the left hydraulic system, electrically engaged by the nosewheel steering button on the control stick, and directionally controlled by the rudder pedals. Retraction and extension time of the gear was approximately 5 seconds. The main gear were held in retracted position by individual uplocks which were hydraulically actuated. The nose gear uplock was contained within the gear drag-brace mechanism. All gear were locked in the down position by spring-loaded overcenter downlocks.

## Landing Gear Alternate Extension

The alternate release D-handle to the left of the instrument panel permitted gear extension with the landing gear control lever up or down. Pulling the handle operated a cable system which opened all the uplocks for gear and opened the main gear emergency power valve, which allowed the gear to free fall. The main gear was forced to the down position by right system hydraulic pressure. The nose gear was lowered by aerodynamic loads and locked by springs.

## Emergency Wheel Brakes

Pulling the emergency brake lever on the left console opened an emergency brake power valve which allowed pressurized hydraulic oil from the emergency brake accumulator to flow to the brake metering valve. Brakes were then controlled in the normal manner by the brake pedals. Approximately 10 full pedal applications were available. Anti-skid control was inoperative when using emergency braking.

## Anti-Skid System

The anti-skid control system provided brake pressure control to prevent prolonged wheel skids, tire damage, and anti-skid

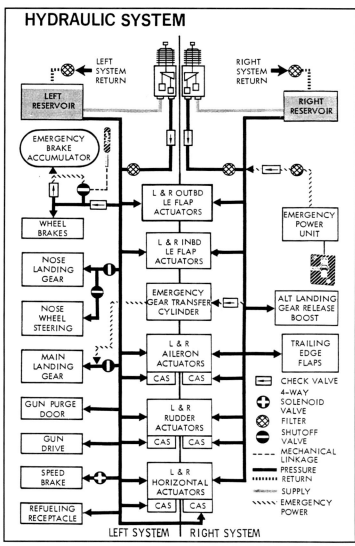

failure detection. The system operated when the landing gear was on the ground at ground speeds above 15 knots. Under 15 knots the aircraft was controlled by manual braking only and the anti-skid was inoperative. The system used wheel deceleration to detect skids; skids on one wheel caused brake pressure to be modulated to both wheels for maximum directional control.

## Flight Control System
Ailerons, horizontal tail, and rudders were positioned by dual-hydraulic closed-loop actuators. All primary flight control surfaces utilized control augmentation system (CAS). CAS hydraulic actuators were integrally attached to the control surface hydraulic actuators. If either hydraulic system malfunctioned, hydraulic power to the flight control system continued to be available. If both hydraulic systems failed, as in dual-engine flameout, EPU operation powered the flight controls. Both rudders, both horizontal tails, and the right aileron were available under this condition.

## Pitch Control
Pitch control was through conventional cable and push-pull rods to the hydraulically-actuated, all-movable horizontal tail which was capable of operation with CAS on or off. CAS provided relatively constant stick force per-G throughout the flight envelope as a programmed response to airspeed and altitude based on data computed by the DADC. Control to the surface actuator was through a dual-mechanical system. Left and right horizontal tail surfaces were independently controlled to permit differential deflection for rolling tail control.

## Roll Control
Ailerons were controlled by a direct electrical (DE) control signal from the control stick to the aileron CAS actuators and a model-following roll rate command system. The DE control of the roll axis provided a signal from the control stick through a function generator. The CAS provided a roll rate gyro signal and a signal from the control stick through the function generator to an electronic model. The difference between the latter two CAS signals provided an error signal. The gain of this error signal was scheduled as a function of compressible dynamic pressure. The DE and CAS were used to drive the respective left and right CAS actuators and actuator models. The input command to each actuator was adjusted by two functions supplied by the DADC which controlled each side of limiters. These limiters CAS actuator travel, and in turn, aileron deflection. Maximum travel was 35 degrees up and 25 degrees down.

## Yaw Control
Dual hydraulically-powered rudders were operated by a single cable system connected from the rudder pedal control mechanism to the aft cable quadrant. This quadrant operated left and right rudder servo valves through load limiters and push-

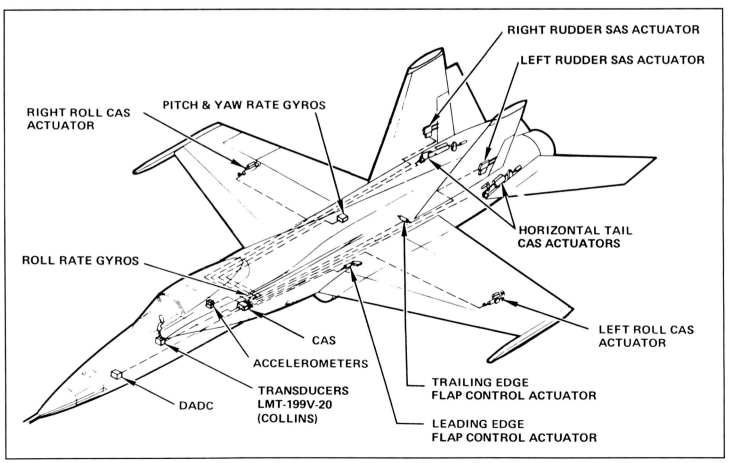

rod linkages. A centering spring attached to the servo linkage centered the rudder should the linkage become disconnected. Each rudder actuator was a dual hydraulic actuator with an integrated CAS hydraulic actuator. Yaw control augmentation consisted of a yaw rate gyro, a lateral accelerometer signal, and a signal from the product of angle of attack and roll rate gyro signals. These signals plus dynamic pressure inputs from the DADC were processed by the CAS system to determine rudder positioning during flight. Failure of the DADC switched off the schedule portion of the yaw rate gain leaving it a nominal value, and would set the lateral accelerometer and the angle of attack/roll rate signals to zero. Aileron rudder interconnect (ARI), and stick/rudder interconnect also provided signals from each aileron and lateral control stick to the CAS actuators. These signals were modified by DADC functions of angle of attack and Mach number. In case of DADC failure, a back-up system replaced the DADC functions with functions of horizontal tail and trailing edge flaps. Maximum rudder deflection in manual mode was 30 degrees either side of neutral with landing gear down and hinge limited to 10 degrees with gear up. In CAS mode, maximum rudder deflection was 25 degrees either side of neutral; however, the amount of deflection during flight was a function of dynamic pressure force (q) on the rudder surface and varied with airspeed, altitude, and angle of attack.

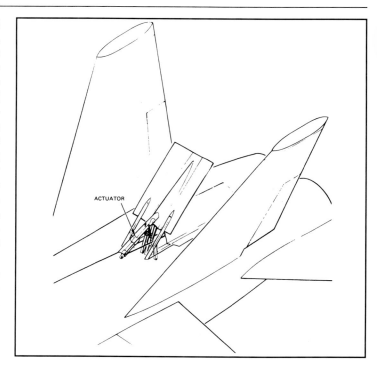

**Flight Control Augmentation System**
The CAS automatically dampened pitch, roll, and yaw oscillations during flight to provide smooth control inputs programmed to allow safe and stable flight through all flight regimes.

**Wing Flap System**
The leading edge and trailing edge flaps were used in all phases of flight below 450 KIAS and were capable of automatic operation with DADC control as maneuvering flaps, deployed as a function of angle of attack and Mach number. The leading edge flaps were powered by both hydraulic systems; trailing edge flaps by the right system only. In the event of right hydraulic system failure, the trailing edge flaps locked in a faired position. Mechanical interconnect, between left and right servo valves assure symmetric flap control. The trailing edge flaps were mechanically interconnected with the horizontal tail to minimize trim changes. An asymmetric detection system shut down the flap drive system if normal asymmetry limit was exceeded. During normal system flap operation, the leading edge and trailing edge flaps could not be independently operated.

**Speed Brake**
The electrically-controlled variable-position type speed brake, located on the upper aft fuselage between the vertical tails, was controlled by the speed brake switch on the master throttle and was powered by the left hydraulic system. Full extension was 60 degrees. At high airspeed, air loads prevented full extension of the speed brake. The speed brake on No 2. (72-1570) was removed to allow installation of a spin recovery parachute during a spin testing.

## COCKPIT

### Cockpit Geometry
The cockpit geometry incorporated several changes from the standard cockpit as follows; the heel rest line was elevated two inches; the seat back angle was 18 degrees instead of 13 degrees, and the seat pan angle was 18 degrees instead of 6 degrees. These changes raise the legs so that the knee was above the hip, with the result that leg blood tends to flow from the knee to the hip under g load, rather than draining to the feet; thus total body blood, being on a level closer to the heart, was more readily pumped to the eyes. The seat back angle was increased moderately to provide pilot comfort with the raised foot position, and to allow excellent aft visibility.

Other cockpit features for increasing g tolerance were:

(1) a control stick with position, deflection, and height optimized for flying with the right arm supported on the thigh,
(2) a foldout arm rest (easily positioned for pilot preference) aft of the throttle to support the left elbow and provide balanced support for precision power adjustments,
(3) handgrips to improve movement in the cockpit at high g,
(4) lateral restraint provided by a contoured back cushion and a standard PCU-15 USAF harness.

### Cockpit Arrangement
The instrument panel incorporated the basic "T" of flight instruments. To the right of the "T" was a set of tape engine instruments. High priority instruments were at the top of the panel for minimum eye movement during combat, armament controls on the left for ready access, and angle of attack and

The cockpit layout and instrument panel were designed to ease the pilot's work load. High priority instruments were at the top of the panel for minimum eye movement during aerial engagement. Armament controls were on the left for easy access, with angle of attack, accelerometer and fuel on the right. (Northrop)

accelerometer on the right. The YF-17 incorporated an advanced-type throttle for twin engine aircraft. The throttle consisted of a single master lever which could control both engines. Oxygen for the pilot was supplied from a 5-liter liquid oxygen converter through the console-mounted regulator. The ejection seat incorporated a bail-out oxygen container. The standard anti-g valve was located in the aft portion of the left console.

## Ejection Seat

The SIIIS-3 ballistic-initiated, rocket powered, DART and drogue parachute stabilized ejection seat provided zero-zero and high altitude/airspeed escape capability. The man/seat connections were designed for use of the PGU-15/P pilot torso harness. Incorporated into the seat was a modified survival kit containing a survival package, emergency locator beacon (AN/URT-33), emergency oxygen bottle, safety belt, and contoured seat pad. A non-adjustable headrest contained the main parachute canopy and risers. Barometric and airspeed sensors, mounted on the right side of the seat structure, determined system mode of operation. The modified G-9 canopy parachute was equipped with a pilot chute and ballistic spreader. A drogue chute was housed in a container in the rear of the headrest. The seat featured a ground safety control handle prominently located on the right side of the seat to arm or safe the seat. Canopy piercers above the aircraft seat supports were designed to pierce upward and then rotate rearward to initiate fractures in the canopy if it failed to jettison. A "spanker" plate across the top of the headrest propagated canopy glass fractures as the seat rose during ejection.

A D-ring handle on the front of the seat bucket between the pilot's legs was used to initiate ejection. An emergency release handle on the right panel was used for manual override of the automatic sequencing systems to release the drogue and main parachute after ejection.

## EJECTION SEAT OPERATION

The following table summarizes the events and times for the four operating modes of seat ejection from initiation until the parachute container opened. Time for parachute container opening depended upon airspeed and altitude at the time of escape initiation. Upon parachute line stretch the spreader gun fired and the drag was sufficient to fire the man/seat release initiator. This severed the shoulder retention reel straps and released the survival kit allowing the man and seat to separate. The four operating modes during seat ejection sequence were:

Mode 1
Low Speed        Below 225 +/-20 kt
Low Altitude     Below 7000 ft.

Mode 2
High Speed       Above 225 +/-20 kt
Low Altitude     Below 7000 ft.

Mode 3
Intermediate Altitude    Between 7000 ft and 14,000 ft.

Mode 4
High Altitude    Above 14,000 ft.

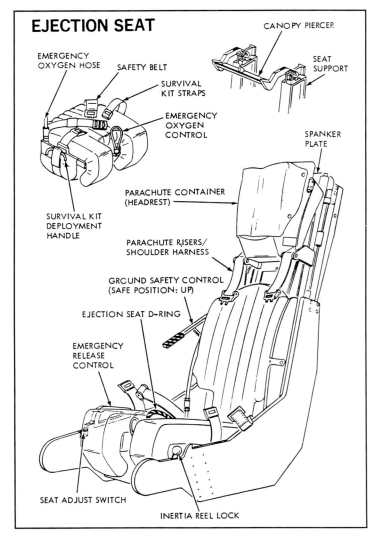

## Canopy

The one-piece jettisonable canopy was electrically powered, and automatically locked and unlocked during normal operation. The canopy jettison T-handle, to the right of the instrument panel, when pulled, would jettison the canopy. Automatic jettison occurred during seat ejection. A canopy breaker tool on the right console could be used to break the canopy glass if all canopy opening methods fail. Two D-handles, within an access door on each fuselage side, were provided for emergency jettison from outside the cockpit.

## Digital Air Data Computer

The digital air data computer (DADC) provided analog-to-digital conversion, air data computation, and auxiliary functions which included monitoring of flight test data and system failure.

### Angle of Attack System

The angle of attack (AOA) system provided a display indication of optimum AOA for landing approach and flight modes of operation. The system consisted of left and right AOA probe type transmitters on each side of the fuselage forward of the cockpit, a side slip angle transmitter on the lower fuselage centerline below the cockpit, an electronics unit in the avionics compartment, an approach indexer light on the HUD gunsight, and an AOA indicator on the instrument panel. Electric heaters in the transmitters were provided for anti-ice protection. The heaters were energized when the weight of the aircraft was off the landing gear.

### Communication and Navigation Equipment

#### UHF RADIO

The AN/ARC-150(V)9 UHF radio on the left console provided transmission and reception within the 225.000 to 399.975 MHz range with voice, tone, and ADF controls. Twenty preset channels plus guard channel (243.000 MHz) could be selected. Manual selection of any frequency (spaced 25 kHz apart) could be accomplished without disturbing the preset channels by use of the frequency selector knobs. An antenna switch on the left console provided L, R, or AUTO antenna selection.

#### INTERCOM

The AN/AIC-18 intercom system provided headset amplification for the UHF radio, IFF, TACAN, landing gear audible warning, AIM-9E missile tone, and cockpit-to-ground crew communication. There were no controls provided for system operation. Individual volume controls on associated equipment were used to control audio level.

#### IFF/SIF

The AN/APX-101 IFF/SIF transponder system, controlled by a standard type C-6280 control panel on the right console, provided automatic identification pulse-coded response to interrogations of ground and airborne radar stations. Operation was possible in any one of three pulse-coded modes, as selected on the control panel, with capability of IDENT (Identification) position and emergency identification. Mode 1 (Security Ident), Mode 2 (Personal Ident), and Mode 3/A (Air Traffic Ident) were operational. Mode 4 (Security Ident-Classified) and Mode C (Altitude Report) were inactive. Modes 1 and 3/A were capable of manual variable selection of required codes in flight. Mode 2 code was preset into the system before flight. Emergency mode of operation could be manually selected on the control panel.

#### TACAN

The ANS-100(V6A) TACAN system provided bearing, distance, course deviation, aural and visual channel indication, and transponds to other aircraft interrogations.

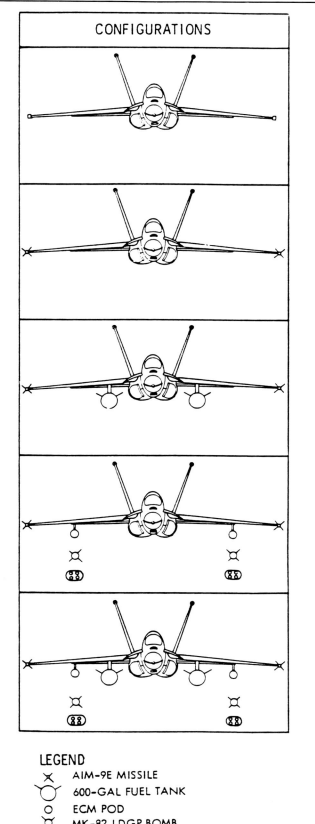

## INERTIAL NAVIGATION SYSTEM

The Litton LN-33 self-contained inertial navigation system (INS) continuously determined aircraft position by sensing accelerations for erection of gyros referenced to true north, and provided precision attitude/heading reference and navigational display information. System components were the computer and inertial platform (inertial navigation unit INU) in the fuselage aft avionics compartment, and the control display unit (CDU) on the right console. Geographic position data programmed into the CDU provided magnetic heading for the TACAN and HSI, and true heading for the HUD sight. Pitch, roll, and attitude warning data was supplied to the attitude indicator. Pitch and roll data was also supplied to the HUD sight. The system was capable of storing the present position (point-of-departure) and nine pre-selected geographic navigation points in the CDU. These navigation points could be changed either on the ground or in flight. Visual update of enroute positions could be accomplished in flight. Digital readout data was displayed on the CDU.

## WEAPONS SYSTEM

Basic armament consisted of the 20mm Vulcan gun in the nose and two AIM-9E type missiles on the wingtips. The fire control system consisted of a range-only radar and a head-up display (HUD) sight system. External stores were carried on four wing pylons. The pylon bomb racks contain electrically initiated impulse cartridges to force eject stores. There was no provision for fuze arming. The master armament switch, on the instrument panel, must be positioned at ARM to fire the gun or missiles, and release external stores.

### Gun System M61A1

The M61A1, 20mm Vulcan gun system was pallet mounted as a unit on the centerline in the nose section. The complete system consisted of the rotating cluster of six gun barrels, the modified ammunition drum with a capacity of 500 rounds of linkless 20mm ammunition, the feed chutes, and the hydraulically powered drive. Access for gun servicing was provided through doors in the bottom and right side of the fuselage. The gun was electrically controlled and fired. Empty cases and cleared rounds were returned to storage in the drum. Although capable of high (6000) and low (4000) rounds-per-minute rate of fire, the gun was electrically wired to provide only high rate continuous fire. A shot port was provided through the upper nose section skin. During firing, a deflector at the muzzle and deflected the gun gas away from the aircraft. Gun breech purge provided air circulation throughout the gun housing and ammunition transfer unit. Air was drawn through by an ejector pump system using engine bleed air. The gun compartment was also scavenged by ram air directed over the gun components.

### Pylons

The inboard non-jettisonable pylons contain MAU-12C/A bomb racks with both 14- and 30-inch hook spacing. The outboard non-jettisonable pylons contain MAU-50/A bomb racks with 14-inch hook spacing. The outboard pylon armament station selector/indicator switches on the instrument panel were used to select normal release only of external stores loaded on these pylons.

### Wingtip Launchers

The wingtip launchers were equipped to carry the AIM-9 series missiles.

## WEAPON SYSTEM CONTROLS

The armament control panel was located on the left side of the upper instrument panel. This panel contained the master arm switch, the gun/camera switch, missile volume control, and pylon/wingtip selector switches.

A capability to selectively launch AIM-9 missiles was provided. Left, right or both wingtip positions could be selected by actuating the switch for that location. The switch armed the missile circuitry and activated the missile guidance and control system. Acquisition of a target by the missile caused

The M61A1 20mm gun assembly including the ammo drum which held 500 rounds was installed in the nose from below, allowing easy access for repair and replacement of the gun. (Northrop)

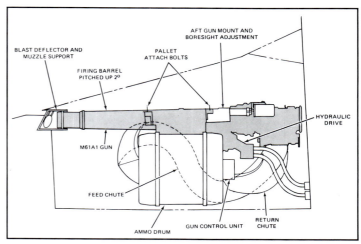

a tone to be generated in the missile which was fed to the pilot's headset. The pilot could adjust the volume of the tone with the MISSILE VOLUME control on the armament control panel. When the pilot heard the tone, he could elect to uncage the missile seeker head if he determined target lead angle was required. The missile was uncaged by actuating the missile uncage switch located on the left hand power control lever. The missile seeker head returned to the caged position upon release of the switch. Whether caged or uncaged, the missile tone indicated that missile guidance lockon had been accomplished, and the missile could be launched by pressing the missile button on the control stick grip.

Gun controls consisted of a three-position gun-camera switch on the armament panel and a conventional trigger switch on the control stick. The gun-camera switch on the panel had a central OFF position and could be positioned to energize both the camera and guns simultaneously, or the gun camera only. The first detent position of the trigger switch actuated the gun camera and purge system; the second detent position of the trigger switch fired the guns.

Emergency jettison from the wing stations was accomplished by pushing the emergency all-jettison button.

### Head-Up Display Gunsight System

The head-up display (HUD) gunsight system consisted of a display unit (DU) centered above the instrument panel and an acceleration and rate converter (ARC) unit in the aft avionics compartment. Symbolic display of data was projected on the combining glass from a cathode ray tube within the DU and appeared superimposed on the forward view through the windshield. A fire control computer within the DU provided solutions to computations based on digital inputs of range, range rate, true airspeed, angle of attack, angle of slip, relative air density; and motion inputs of vertical and lateral acceleration, roll and pitch attitude; and angular rates of turn in roll, pitch and yaw as required for positioning of the symbology. In addition, the system provided boresight and self-test (BIT) displays. Input data was received from the Fire Control Radar, DADC, INS, Armament Controls, and CAS.

## FIRE CONTROL RADAR

The X-band radar provided range and range rate only data to the HUD sight system. Components were the flush surface wave antenna in the bottom of the forward nose section and the transmitter-receiver in the forward nose section. In the ON position, the radar would search, detect, lock on, and range track in accordance with selected operation of the sight system controls. Acquisition range was approximately 5000 ft. Search range was limited to 6000 ft to avoid ground clutter.

## FLIGHT TEST INSTRUMENTATION

The installed airborne data acquisition equipment consisted of digital data systems using serial pulse-code modulation (PCM) recorded on magnetic tape and telemetered. High frequency data was recorded in analog form. Cameras could be installed to provide specific functions. The equipment consisted of the nose boom, emergency power unit, C-band radar beacon, fire extinguishing system, and spin recovery parachute. The instrumentation consisted of the PCM system, tape recorder, data processor, and associated units.

## PROTOTYPE AIRCRAFT DIFFERENCES

Prototype No.1 (72-1569) and Prototype No.2 (72-1570) were equipped with different cockpit controls/indicators to facilitate flight test events. This instrumentation, in some instances, temporarily replaced previously installed equipment. During scheduled spin flight tests of 72-1570, a third battery was installed in lieu of the TACAN receiver-transmitter in the avionics equipment bay. This battery provided emergency backup power for spin chute deployment and jettison, UHF radio, INS, telemetry data output, and data recording.

## TEST CONFIGURATIONS

The basic aircraft configuration was wingtip launcher rails and no pylons. Launcher rails were equipped to carry AIM-9E missiles. Outboard pylons (when installed) could carry dummy ECM pods, inert MK-82 LDGP bombs, or the SUU-20A/A bomb/rocket dispenser loaded with inert 2.75-inch FFAR rockets and inert practice bombs. Each inboard pylon (when installed) could carry a 600-gallon fuel tank. On 72-1570, a dummy centerline pylon with a 300-gallon non-feeding fuel tank could be carried to test drag characteristics. The M61A1 gun was installed in 72-1569. Ballast equivalent to the gun was installed in 72-1570. During spin tests with 72-1570, the speed brake installation was removed and a spin recovery parachute and controls was installed.

## SPIN RECOVERY PARACHUTE

The spin recovery parachute was installed in place of the speed brake system on the upper aft surface between the vertical stabilizers on the second aircraft (72-1570) only. The parachute had a flat diameter of 26 feet and riser length (shackle to parachute canopy) of 100 feet. The parachute riser incorporated thermal and abrasion protection. The parachute riser was attached to the airframe at a lock mechanism controlled from the cockpit and actuated by an electromechanical actuator powered from the primary battery bus. Four deployments were made of the emergency spin recovery parachute to demonstrate system operation and to obtain parachute deployment data. One ground deployment at 65 KIAS and three inflight deployments of 120, 165, and 190 KIAS were performed. During the stall and post-stall tests, for which the spin chute was installed, no deployments were necessary.

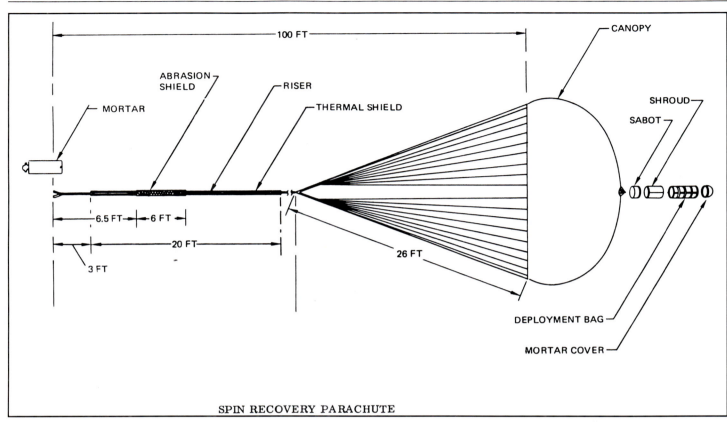

SPIN RECOVERY PARACHUTE

**This view shows the temporary spin recovery parachute installation on the fuselage top between the twin vertical tails which replaced the speed brake on 72-1570. (Mick Roth)**

*Also from the publisher*

*Other books by Don Logan*

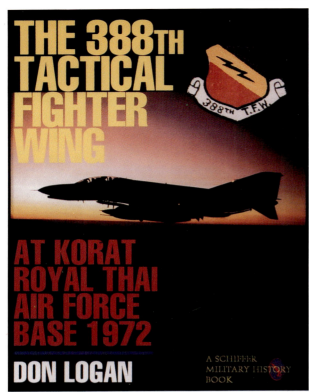

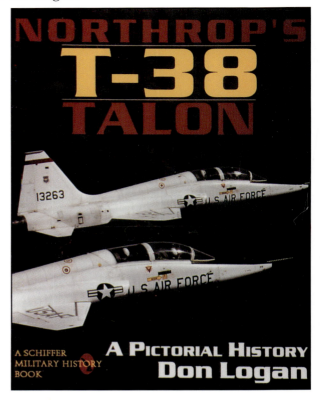

## THE 388TH TACTICAL FIGHTER WING AT KORAT ROYAL THAI AIR FORCE BASE 1972
### Don Logan

This new book covers the 388th TFW; a Composite Wing based at Korat RTAFB, Thailand, consisting of fighters, Wild Weasel aircraft, airborne jamming aircraft and AWACS aircraft. The author flew 133 combat missions in Southeast Asia in 1972, and was assigned to the 469th TFS, one of the two F-4E squadrons of the 388th TFW. The book discusses in detail the Wing, the Squadrons and the aircraft they flew: the F-4, F-105G Wild Weasel, A-7D, EB-66, EC-121, and C-130. Also covered are the mission types, as well as operations of the Wing during the Linebacker Campaign over North Vietnam. Narratives of all the 388th MiG kills and aircraft losses during 1972 are included. The book contains over 170 color and black and white photographs taken by the author, as well as theatre maps. A selection of official and unoffical flight suit patches is also included. Don Logan is also the author of *Rockwell B-1B: SAC's Last Bomber*, and *Northrop's T-38 Talon: A Pictorial History* (both titles are available from Schiffer Publishing Ltd.).

Size: 8 1/2" x 11"  over 170 color and b/w photographs, maps
128 pages, hard cover
ISBN: 0-88740-798-6                                      $29.95

## NORTHROP'S T-38 TALON A PICTORIAL HISTORY
### Don Logan

This is the story of the most successful pilot training jet ever produced: the Northrop T-38 Talon. The history of the aircraft is broken down by the roles it has played in over thirty years of service including development and testing, pilot training, flight test support, NASA program support, air combat aggressor, aerial target, Thunderbird-USAF air demonstration team aircraft, companion trainer, and civilian test support. All units flying the T-38, their markings and paint schemes are covered in over 300 color photographs – including a chart of the colors used listing Federal Standard (FS) color numbers. Don Logan is also the author of *Rockwell B-1B: SAC's Last Bomber*, and *The 388th Tactical Fighter Wing: At Korat Royal Thai Air Force Base 1972* (both titles are available from Schiffer Publishing Ltd.).

Size: 8 1/2" x 11"  over 300 color photographs
152 pages, soft cover
ISBN: 0-88740-800-1                                      $24.95